Standardized Work for Noncyclical Processes

Standardized Work for Noncyclical Processes

Joseph Niederstadt

CRC Press
Taylor & Francis Group
Boca Raton   London   New York

CRC Press is an imprint of the
Taylor & Francis Group, an **informa** business

A PRODUCTIVITY PRESS BOOK

Productivity Press Taylor
& Francis Group 711
Third Avenue New York,
NY 10017

© 2010 by Taylor and Francis Group, LLC
Productivity Press is an imprint of Taylor & Francis Group, an Informa business

No claim to original U.S. Government works

10 9 8 7 6 5 4 3 2 1

International Standard Book Number: 978-1-4398-2550-1 (Paperback)

Library of Congress Cataloging-in-Publication Data

Niederstadt, Joseph.
   Standardized work for noncyclical processes / Joseph Niederstadt.
     p. cm.
   Includes index.
   ISBN 978-1-4398-2550-1
   1. Assembly-line methods. 2. Automobile industry and trade. I. Title.

TS178.4.N54 2010
658.5'1--dc22   2010006506

Visit the Taylor & Francis Web site at
http://www.taylorandfrancis.com

and the Productivity Press Web site at
http://www.productivitypress.com

Contents

Supplementary Resources Disclaimer

Additional resources were previously made available for this title on CD. However, as CD has become a less accessible format, all resources have been moved to a more convenient online download option.

You can find these resources available here: www.routledge.com/9781439825501

Please note: Where this title mentions the associated disc, please use the downloadable resources instead.

Preface

I have had the opportunity in my working years to be taught the whys and hows of Standardized Work by some of the best within Chevrolet Motor Division, Delphi Corporation, and Toyota. This book represents a compilation of my 33+ years of manufacturing experience with Standardized Work. It all began when I was an assembly line operator at a global automotive manufacturer, and I didn't even know what Standard Work was at that point.

My first day as an assembly line operator began with an orientation of rules, regulations, benefits, UAW introduction, and a mound of paperwork. I had no clue what it meant to work in a large manufacturing environment and didn't care; I had hit it big by landing a job in General Motors.

After the orientation, I was led to an assembly line. As I observed the line running, I began thinking, "I'll never keep up with that line. It is too fast and complicated. I'll get fired my first day. What have I gotten myself into?" I was placed in one of the 17 positions on the line with another new operator. Together we were instructed verbally on how to complete our work. They put two of us together so we could learn the job, or so I thought. The real reason they put the two of us together was so we would not slow down production because production output was the only goal in which anyone was interested.

Our line was rated to run a maximum of 600 parts per hour. If you missed a part or slowed the line, a horn would sound and a line leader or supervisor would come down and ask why or make the comment: "You too stupid to keep up?" The seasoned workers usually stayed away from us because we were new. I found the whole thing to be a humbling and humiliating experience. Eventually the other person would be removed and you were expected to do the job by yourself. At first I was nervous and sweated a lot. Then I fell into a rhythm and didn't experience any problems. In fact, I figured out how to do it faster so that I had some free seconds on my hands. It had never occurred to me at that time that somewhere a process or industrial engineer had created standard work for this process so it could be accomplished with ease in the given time. This was my first lesson in Standard Work.

Eventually, I began to get bored with doing the same routine every day. Once, I offered an improvement idea to my supervisor who promptly told me to "shut up and build parts." I found this attitude to be ignorant and irritating. Over time, I was given an opportunity to learn more jobs of the line and work my way up the hierarchy of manufacturing life. As a supervisor, I learned that every manufacturing machine in the plant had an expected output per hour and that every manufacturing job in the plant had been designed to meet that rate. So, we would document our production's performance every hour. The boss would come around and check your production counts each hour. If any hour was below the expected rate, you had better have a good reason and have it documented as well. This was my second lesson in Standard Work.

I was formally introduced to the Standard Work process in the late 1980s. People from the industrial engineering group began showing me how the process works and what documents were required. Specifically, the industrial engineering team at Chevrolet Saginaw Manufacturing spent a great deal of time answering all of my questions and were key in launching my learning experience of Standard Work. My plant manager and area manager allowed me the time and resources to continue my training in this as well as other areas of Lean tools. Also, a man by the name of Bill Butterfield, who understood my value to the organization, believed in me and utilized my abilities to the fullest extent. I am thankful to all of them.

In the early 1990s, I was further exposed to Standard Work during the time that Toyota was mentoring Delphi on implementing Lean into the factory, transitioning from Mass to Lean thinking and process design. One Toyota sensei was my mentor for understanding the Toyota way of implementing Cyclical Standard Work and another was my mentor for Noncyclical Standard Work.

Between then and now I have been applying these tools to a variety of manufacturing and administrative processes. I have never forgotten my roots, my experiences as an operator, and I hope I never do. One of the best things I learned from the people of Toyota is that "there is no such thing as a bad operator, only bad processes." That was my third lesson in Standard Work.

Later in my career I was introduced to Standard Work for noncyclical processes by one of my Toyota sensei. It did not take long for me to realize the potential for this application. I cannot tell you how many times as a supervisor of maintenance, shipping and receiving areas, or quality auditors I would scratch my head and think there has to be a better way to help these people perform their tasks in a more efficient and logical manner. There were days when I could see the frustration on their faces because they wanted to do a good job, but were faced with roadblocks that prevented them from taking pride in their work.

One of the best tools we have is our eyes. The power of observation prior to and even after standardization is immense. Many times we are in too much of a hurry to make proper observations or don't realize the power of observation—both of these are a mistake. We must go to the *Gemba* (Japanese for where the action takes place) and see what is happening. Never take anyone's word for how a process works or make assumptions of how a process works; get out of your office or your chair and go to the *Gemba* and see.

Why Should I?

You may be asking yourself this very question. Why should I (or the organization) develop standardized work for noncyclical processes? My question to you would be: why wouldn't you? First and foremost, we must remember we are on a Lean journey and it is all about continuous improvements everywhere.

We should be asking ourselves this question daily: "What did I do today to make my job better?" If you cannot answer this question with an improvement, then you are in what is known as "firefighting" mode. Simply put, during the course of the work day, we are moving from crisis to crisis and solving it as quickly as possible to keep the machine going and producing the customer orders for the day. Although this creates daily "heroes," it does not provide for a culture of daily incremental improvements. I do not use the term *heroes* with any disrespect. These people do everything they can, every day circumventing roadblocks because of poor processes to make sure the job gets done. However, we need to have a culture of daily continuous improvements to reduce costs and stay competitive. Your competition, however, hopes you do nothing; that is the best thing that can happen for them. They will take your customers, your jobs, and your quality of life, but only if you allow it to happen by doing what they hope for, nothing.

Based on my experiences, this is the situation I generally find for noncyclical processes in a business. The entire focus for the application of Lean tools has been in the productive manufacturing environment. That is a good place to start. But, what about all the rest of the activities in a business? Are we just supposed to sit back and say, "Well, that is the way it is with fork truck drivers, material handlers, mechanics, skilled trades quality auditors, SGA, etc." Of course not. These functions are just as much a part of our business as making product for the customer. They have a significant impact on our cost of doing business and to ignore them is a major missed opportunity for you.

In the world of the cyclic operator, you would not accept a work load of only 20% to 28% would you? Of course not. Then why is it acceptable for the noncyclical processes in a business? Again, based on my experience, when we first look at a noncyclical process, this is exactly what we find, usually these specific percentages of a work load. Surprising information, wouldn't you agree?

The process of developing Standardized Work for noncyclical process will help you make those sustainable incremental improvements to any noncyclical process you have in your organization as well as improve the flow of work throughout the entire value stream. It will help considerably in getting your organization out of the firefighting mode and into the continuous improvement practice.

Acknowledgments

I would like to thank my wife, Terri, for her continued support and tolerance over the past 33 years of my life in the manufacturing environment and for providing me the encouragement to bring what was once just a comment, "You ought to write a book about what you do and know," to fruition. Also to Tim, Robyn, Tom, and Frank who constantly asked, "When are you going to write that book," or "How is that book coming along?" That put me in a position where I had to do something. I would also like to recognize my Mom, who from the time I could understand, drove into my brain that I could accomplish anything I set my mind to. I would also like to thank Pat Straney, Merrill Reinig, Mike Minadu, Tim Maxwell, and Ken Baybeck and Bill Butterfield who helped me throughout my Lean journey more than they realize, as well as the thousands of people I have interacted with over the past 33 years of my career. There have been so many people that have had an influence on my life that I could not begin to name them all. Some people had a positive influence, some a negative one, but I always learned, and continue to do so, something from both experiences.

This turned out to be a fun project for me. I enjoyed the time at the computer trying to think of how to put my actions into words so people can learn from my experiences. I guess you would have to say this book is one of the items on my "bucket list." I really had fun putting this material together.

Introduction

In the industrial world, we find two different kinds of work: cyclic and noncyclic. Cyclic work is defined as a repeatable process. Noncyclic work is defined as work where the key elements are repeatable, but the sequence in which they occur is not repeatable. Cyclic and noncyclic job processes are not the same. We must separate these processes to maximize effectiveness and efficiency. Would you like your company to maximize the efficiencies of indirect labor assignments, such as multiple machine job setter, material movement, maintenance, stockers, order pickers, and quality auditors? These are known as noncyclical types of activity and are generally labeled as *indirect labor*. These positions are required to support the value adder in the manufacturing work environment. Have you ever observed a fork truck driver or material handler when their vehicle is moving, but there is nothing on the fork racks or vehicle bed? Have you noticed people doing a lot of walking or waiting? Have you seen people walking or driving more than working? If you can answer, "yes," to any of these questions, noncyclical Standardized Work can help you.

Most companies have some form of work instructions for indirect labor positions, but they stop there. It appears to be a common belief that waste associated with these positions is a normal condition. This simply is not true. Remember, without standardization, stabilization, and the elimination of variation of the process, you cannot make continuous improvements. The same is true for noncyclical processes.

Where there is no standard, there is no improvement.

Taichi Ohno
Toyota

Over the years, I have heard many reasons companies use to ignore this opportunity to remove waste:

1. This is the way we have always done it.
2. I didn't know there was anything I could do.
3. My people know how to do their jobs.

These are terrible excuses for not doing the right thing for your operators, department, and your company. This type of thinking can kill your competitive edge in cost, set you up to lose business due to poor quality, cause you to have a huge amount of capital tied up in inventories that are in the wrong place at the wrong time, and have a negative impact on your productivity due to waste built into the process.

One of the fundamental principles of Lean is to have stable and reliable processes. I am frequently asked, "How do we apply typical Standardized Work concepts to the noncyclic-type job?" The purpose of this book is to help you identify the waste, create Standardized Work for your noncyclical processes, and make sustainable, continuous improvements. It will provide you with definitions of documents, how to use the documents, examples of documents, and explain a step-by-step instruction on how to collect and interpret the data and which Standardized Work documents should be used.

What makes this book unique? For starters, I am not an engineer. I am just a person who started as an operator, worked my way through the ranks, took an interest in this subject, and learned how to apply this tool. More importantly, I have taught this approach in more than 30 factories globally. It has been successfully applied to hundreds of various noncyclical manufacturing and nonmanufacturing processes around the world. Throughout the book, you will notice references to blank forms as well as various examples of what these forms may look like when they are populated with collected data. On the CD that accompanies this book, for your convenience, you will find full-color versions of many of the figures throughout the book as wells as copies of all the blank forms mentioned. I hope you will find these electronic forms and figures very helpful. Just remember, if I can do it, you can, too, and this has been proven with every application.

Chapter 1

Benefits and Prerequisites

The entire process included in this book is about collecting data to help one make decisions on actions that will increase safety, quality, and productivity while reducing costs. It is purposely laid out in a stair-step approach to allow for time to collect data and collate it into a format that will assist in the next step of the process. Do not try to leap over a step or you will not reap the benefits overall and the changes you make will not be sustained for the long term.

The Benefits

The benefits of implementing noncyclical Standardized Work are actually limitless. As I mentioned in the Introduction, the goal of this process is to increase safety, quality, and productivity while reducing costs. This is accomplished by removing the waste from a noncyclical process in the same manner you would on a cyclical process, and that is by using data, not emotions, to implement sustainable change. This process can be applied not only to noncyclical work that supports manufacturing processes, but also in noncyclical processes in all engineering departments, financial departments, human resources, etc.

The benefits to management can include:

- Providing an efficient and safe work method
- Defining what is the normal state
- An efficient training tool that eliminates guesswork
- Identifying abnormalities and waste

■ Describing and defining a detailed routine that supports building a consistent quality product at a reduced cost
■ Establishing the baseline for continuous improvement

The benefits for an operator can include:

■ Providing an efficient and safe work method
■ An efficient training tool that eliminates guesswork
■ Immediately identifying abnormalities and problems
■ Creation through teamwork and knowledge sharing
■ Ensuring the workload will be balanced between multiple operators

Overall, the main benefits are that noncyclical Standardized Work development:

■ Eliminates the waste of walking, waiting, and other nonvalue-added and unnecessary tasks.
■ Stabilizes cost and, through sustained continuous improvement, lowers cost.
■ Provides the means to effectively utilize the nonvalue-added (but necessary) positions that are required to support manufacturing and run the business.

We will follow several of the basic Lean applications to achieve our goal of reducing and eliminating waste; one of which will be utilizing a Cross-Functional Team. A Cross-Functional Team is comprised of people (but not limited to) from the work group, team leaders, and supervisors from the areas of maintenance, engineering, purchasing, and material departments. Building this foundation is critical for future improvements to continue.

One of the most powerful tools is observation. This simply means going to the area—also known as going to the *Gemba* (the actual place)—to see what is actually happening and documenting what we see. Naturally, we also will be following the Plan–Do–Check–Act (PDCA) cycle. After we have collected our data and see the waste, we will develop a "Plan" to reduce or eliminate the waste, implement or "Do" our plan, let the plan stabilize, and then "Check" (audit) to see if we have achieved the expected result and if it is being sustained. And, finally, "Act" on any abnormalities that are blocking our expected results and bring them back to a standard. However, before we get started, we must first cover the four prerequisites to starting noncyclical processes.

Four Prerequisites

The four prerequisites listed below are actually the base for the planning stage prior to scheduling any activity. With proper planning, your scheduled activity will have better flow and the participants will more fully understand the concepts that are being applied. The participants need to understand the concepts before they become fully engaged in the implementation and sustainment of the improvements that will be applied.

First Prerequisite

Before attempting to learn to implement and apply Standardized Work for *noncyclical* processes, you must have a clear understanding and practical application experience in implementing Standard Work for *cyclical* processes. To make sure we all start at the same place, let's first review the basic definitions for the documents involved.

Work Observation Chart

This document is a drawing (electronic or by hand) of the cell layout, walk, and work pattern of an operator (Figure 1.1). It also shows where Safety, Quality, and Error Proofing concerns, and Work-in-Process (WIP) are located throughout the Operation displayed. In addition to this descriptive information is the use of Safety, Quality, Error Proofing, and WIP symbols—Takt and Cycle Time—as well as process ID. (Takt is the pace at which one of a finished product needs to be made to meet the customer demand. Takt time is calculated by dividing the available operating time by customer demand.) The original document should be kept with the Standard Work document package for the cell in the industrial engineering office or at the supervisor's or team leader's desk. A reference copy also can be posted at the cell or maintained at the supervisor's/team leader's desk.

How to complete a Work Observation Chart:

- **Plant Location:** Enter the correct code for the appropriate plant location.
- **Cell Identification:** Enter the correct code for the cell being studied.

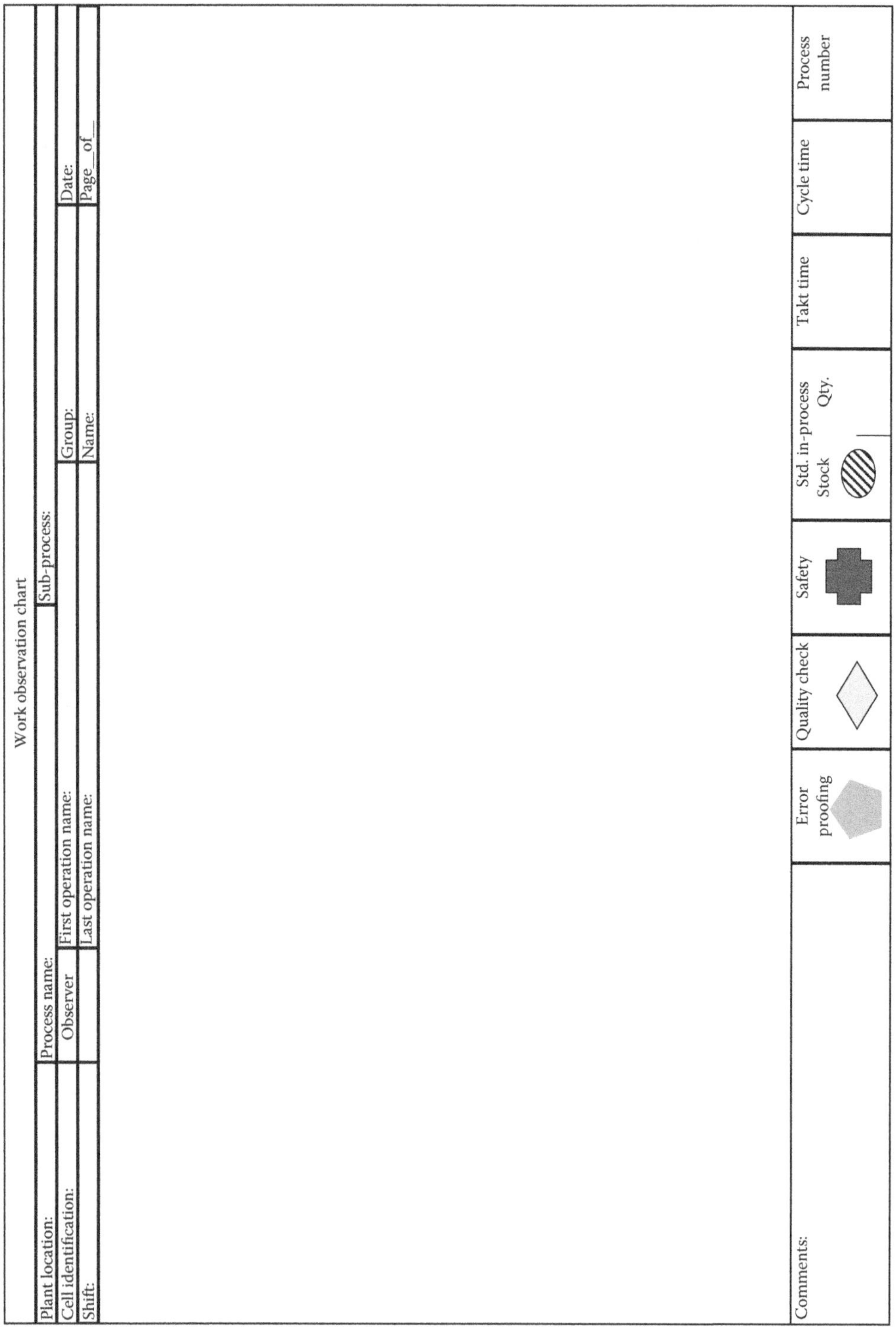

Work observation chart

Plant location:	Process name:		Sub-process:		
Cell identification:	Observer	First operation name:		Group:	Date:
Shift:		Last operation name:		Name:	Page___ of___

| | Error proofing | Quality check | Safety | Std. in-process Stock   Qty. | Takt time | Cycle time | Process number |
| Comments: | | | | | | | |

Figure 1.1   Work observation chart.

- **Shift:** Enter the shift number or identifying character (such as 1st, 2nd, or A, B, etc.).
- **Process:** Enter the main process name.
- **Subprocess:** Enter the exact subprocess name for the area being studied.
- **Group:** Enter the group name.
- **Name:** Enter name of the operator.
- **Date:** Enter the date of the observation.
- **Page:** Enter the page number with indication to the total number of pages.
- Draw the equipment/process layout.
- Enter the machine and/or operation numbers on the drawing.
- Show the working sequence by identifying the first and last sequence with numbers and drawing lines as to the walk pattern of the operator. Observe the operations long enough to determine the repeated (cyclic) activities.

Standard Observation Sheet

This document is used as a Visual reference tool and Auditing tool to verify that the Standard Work is being performed as documented in a cell and is posted in the work area (Figure 1.2). It is not meant or designed to be used as Work Instruction. This document lists, in general, the job elements required to perform the task. It will guide the reader to a Work Instruction-numbered document when more detail is required. It should show the symbols for, but not limited to, safety concerns, quality checks (visual or in-line gages), and error proofing devices at the point of the general job element where applicable. The document also shows the information created for the Standard Work Chart or Sheet. The header provides descriptions of the area defined, revision number, and signoffs for the supervisor and team leaders. This document also should describe and show what personal protection equipment is required for the operator. The original document should be kept with the Standard Work document package for the cell in the industrial engineering office, or supervisor's, or team leader's desk.

How to complete a Standard Work Observation Sheet:

- **Plant Name:** Enter the name for the plant location.
- **Cell Identification:** Enter the identification name or number for the cell.

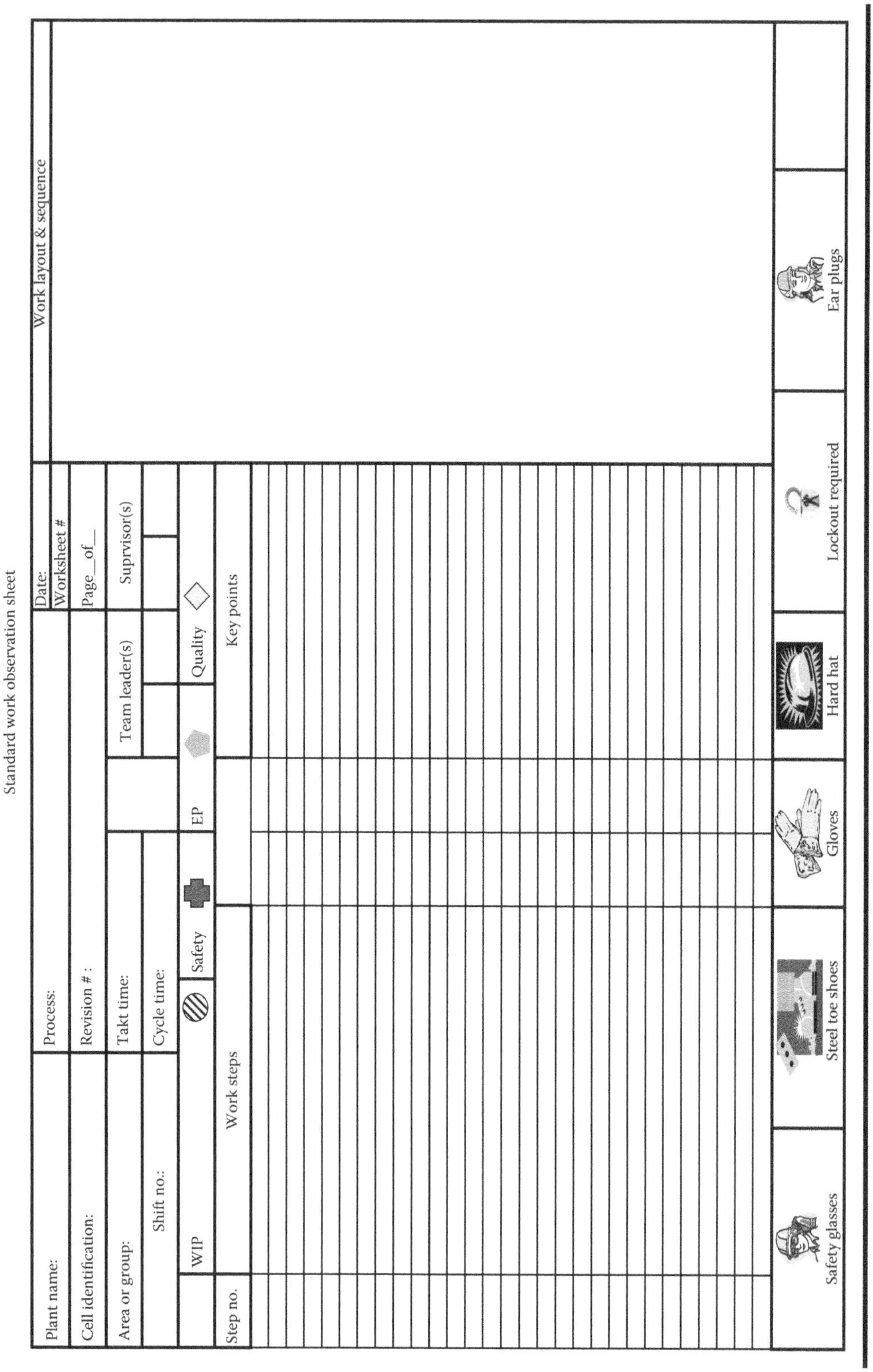

Standard work observation sheet

Plant name:		Process:		Date:	Work layout & sequence
Cell identification:		Revision # :		Worksheet #	
Area or group:		Takt time:	Team leader(s)	Page__of__	
Shift no.:		Cycle time:		Suprvisor(s)	
WIP	〰 Safety	✚	EP	⬠	◇ Quality	
Step no.	Work steps				Key points	

| Safety glasses | Steel toe shoes | Gloves | Hard hat | Lockout required | Ear plugs |

Figure 1.2 Work observation sheet.

- **Group/Area:** Enter the group name or area number.
- **Shift:** Enter the shift number or alpha character (Example 1st, 2nd, or A, B, C, etc.).
- **Process:** Enter the main process name.
- **Revision #:** Enter the latest revision number.
- **Date:** Enter the current date when Standard Work Observation Sheet was completed.
- **Worksheet Number:** Enter the assigned worksheet number.
- **Page:** Enter the page number with relation to the total number of pages.
- **Takt Time:** Reference the Work Observation Chart.
- **Cycle Time:** Reference the Work Observation Chart.
- **Team Leader:** Enter the initials of the team leader(s) for this area.
- **Supervisor:** Enter the initials of the supervisor(s) for this area.
- **Step Number:** Enter the number of the work step in numerical order that the operator performs to complete one full job cycle.
- **Work Steps:** The work steps are the actual work the operator performs to complete one full cycle, and broken down into concise logical elements. Enter in all job elements and use as many pages as necessary to complete the Standard Work Observation Sheet for all job elements.
- The "Key Points" column is used to emphasize the important parts of the job element steps and/or sequence. Enter the key point and what the key point relates to, such as quality, error proofing, safety, etc. (use the symbols where appropriate).

Job Element Data Collection Sheet

This document is a data collection tool used to determine the lowest repeatable time a job element can be performed (Figure 1.3). Once the lowest repeatable time is established, it becomes the standard and the baseline for continuous improvement.

How to complete a Job Element Data Sheet:

- **Operator:** Enter the name of the operator.
- **Process:** Enter the process name.
- **Recorded by:** Enter your name.
- **Date:** Enter the date the data were collected.
- **Work Elements:** Enter the predefined work elements in proper sequence, derived from your observations.

Job Element Data Collection Sheet

Operator:

Date: Recorded   Process
         By: Name:

Steps	Work Element	Start Point	Stop Point	Repetition				Time in Seconds :						Lowest Repeated	Walking	Waiting	Potential Opportunity
				1	2	3	4	5	6	7	8	9	10				
1																	
2																	
3																	
4																	
5																	
6																	
7																	
8																	
9																	
10																	
11																	
12																	
13																	
14																	
15																	
16																	
17																	
18																	
19																	
20																	
21																	
22																	
23																	
24																	
25																	
26																	
27																	
28																	
29																	
30																	

Use for balance chart ⟶ 0

Figure 1.3   Job element data collection sheet.

- **Starting Point–Stopping Point:** Enter a description of what starting point you started your stopwatch and the point at which you stopped the watch.
- **Time:** Enter the time for each element collected by using the stopwatch. Any departure from the work sequence should not be used. Only the times for a properly sequenced job element should be evaluated. If the work is not to standard or proper sequence, the operator will need to be retrained and/or the job element sequence updated.
- **Lowest Repeated Time:** This time will be derived from the lowest, most repeated time across the row of each element. Remember, it is not always the lowest time.

Operator Work Instruction

The Operator Work Instruction document is a detailed teaching and certification tool (Figure 1.4). It should be posted at the operator workstation. It is the standard for training and subsequently used for operator certification on the training matrix. It should describe in detail each job element including, but not limited to, safety concerns, quality checks (visual or in-line gages, error proofing devices, gage calibration, etc.). This document also should describe or show what personal safety equipment is required for the operator. This tool is owned by the cell group and provides flexibility for the plant to include pictures and/or drawings to best explain the intricacies and complexities of a process.

How to complete the Work Instruction Form:

- **Plant Code:** Enter plant location, name, or number.
- **Cell Identification:** Enter the name of the cell or the identification number.
- **Area/Group:** Enter area or group name.
- **Process:** Enter the main process name.
- **Subprocess:** Enter the exact subprocess name for the area.
- **Takt Time:** Enter the time required to meet the customer demand.
- **Cycle Time:** Enter the time it takes to complete one work cycle.
- **Sign-off/Approval:** Enter initials of members on each shift responsible for observing and auditing the standard work as it is being performed. Initials represent understanding and agreement to the content of the current standard for job performance.
- **Date:** Enter the date the sheet is created; represents "current standard."

Operator Work Instructions

Plant Code:	Process:			Date
Cell ID NO.:	Sub-process:			Worksheet #
Area/Group:	Takt Time:	Area Leader	Facility	Area Manager	Page of
Shift No.:	Cycle Time:			

In-Process Stock ◯ | Safety ✚ | Quality ◆ | Error proof ⬠

KEY No.	Work Elements/What to do	Key Points/How to do it (Quality, Safety, Knack)	Sketch Drawing / Documents (Highlight Key Points)

Safety glasses	Steel toe shoes	Gloves	Hard hat	Lockout required	Ear plugs

Figure 1.4 Operator work instruction form.

- **Worksheet Number:** Enter the document identification number and revision number, if any.
- **Page of:** Enter sheet number; use "1 of 5, 2 of 5, etc." depending on how many sheets are used.
- **Reference Document Number:** Enter any other documents tied into this work instruction, such as Process Failure Mode and Effects Analysis (PFMEA), Control Plans, etc.
- **Work Element Number (Key):** Enter numbers for sequential work elements.
- **Work Element Description:** Enter actions necessary to complete operation. Definition of an action is any step or motion that advances the work sequence toward completion of a value-added step. Detailed actions should be omitted. Indicate "what to do."

Work Combination Table

This document is a visual display of operator activity (Figure 1.5). It shows the walk, wait, manual, and machine time as it relates to Takt time of a process or operator work content. It helps to see the waste in a process so that resources can be focused on reducing or eliminating that waste. This original document should be kept with the Standard Work document package for the cell in the industrial engineering office, supervisor's or team leader's desk. A reference copy can be posted at the cell or maintained at the supervisor's/team leader's desk.

How to complete a Standardized Work Combination Chart:

- **Plant Name:** Enter the name or plant location.
- **Cell Name:** Enter the name or address number for the cell.
- **Process Name:** Enter the process name. This represents how the material or parts are processed and the status of their development.
- **Machine Name:** Enter the main machine name.
- **Group/Area:** Enter the group name or the area.
- **Shift:** Enter the shift number or alpha character (Example 1st, 2nd, or A, B, C, or first, second, etc.).
- **Date:** Enter the current date when the Work Combination Table was completed.
- **Cycle Time:** Enter the time from the beginning of cycle to end of the cycle.
- **Takt Time:** Enter the calculated or provided Takt Time.

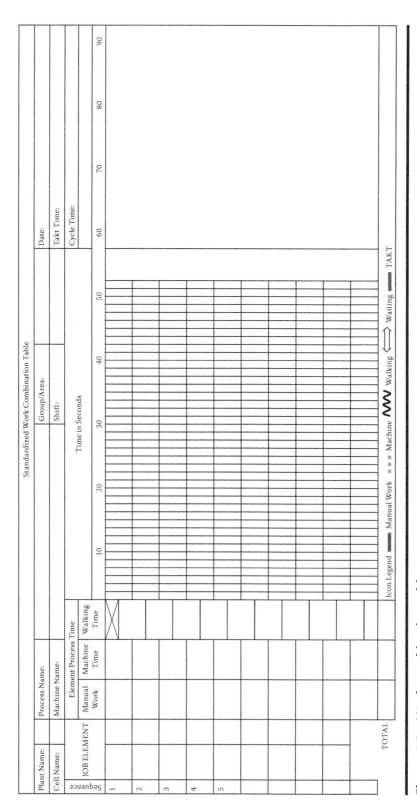

Figure 1.5 Work combination table.

- **Element Process Time:** Enter the work times and machine times you collected on the Job Element Data Sheet and the Production Capacity Sheet. (If there is any doubt of accuracy of the data, actually measure the work time, machine time at the area.) Measure and enter the operator walk time at the site.
- **Waiting Time:** Observed wait time or, if the cycle time is less than Takt Time, enter the difference as wait time.
- **Graph:** Using the icons noted in the legend, complete the graph.

The work balance chart is another standard tool used for developing standardized work for cyclical process. We will be using this tool with a slight twist to aid us in seeing what the data we collect are telling us. This is called the noncyclical yamazumi (long cycle work balance) chart. (Yamazumi literally means "to pile in heaps" and is a tool to achieve line balance with strips of paper or card representing particular tasks.) It is referred to a long cycle balance because it is collecting data over the entire shift period, usually eight hours. Accurate data collection for cyclical processes can normally be achieved with a two-hour process review.

Noncyclical Yamazumi (Long Cycle Work Balance) Chart

This document is a graphic visual display of an operator's or group of operators' daily activity (Figure 1.6 and Figure 1.7). It segregates all the actions of an operator into work that has been defined as nonvalue added, but necessary (for instance, a truck driver is expected to drive with a load of material, not an empty load; a mechanic is expected to be fixing things, not waiting; etc.) and nonvalue add for a noncyclical job. It includes walk and wait (both choice and no choice), and it breaks down the time of a process, operator's or group of operators' daily activities, into graph and data formats. It helps to see the waste in a process so that resources can be focused to reduce or eliminate that waste. This original document should be kept with the Standard Work document package for the cell in the industrial engineering office, supervisor's or team leader's desk. A reference copy can be posted at the cell or maintained at the supervisor's/team leader's desk.

There are two types of master (a base file that needs to be kept as a non-populated file for multiple uses as process reviews occur) yamazumis that will be referenced in this book. One type is when data collection is done by using stopwatches, and the other is when data collection is done by barcode scanning. Each master yamazumi comes with an instructions tab on how to

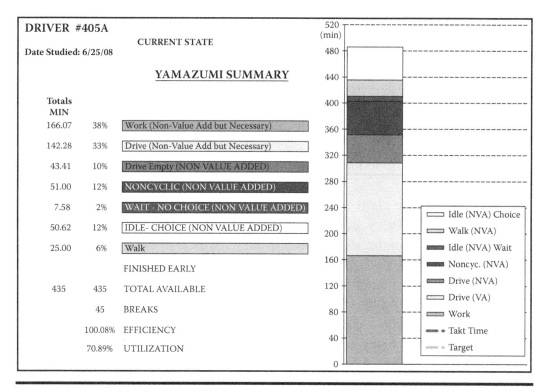

Figure 1.6 Individual yamazumi chart.

use this tool. The yamazumi is a data collection device that utilizes a combination of color-coded stacked bar charts to create a visual display of where an operator's time is being spent throughout the work period. It is used to show you where the waste in a noncyclic process is so that the waste can be reduced or eliminated (see Figure 6.1 and Figure 6.2).

The flow chart in Figure 1.8 helps show the differences in implementing Standard Work for cyclic and noncyclic processes.

Second Prerequisite

You will need to select a specific job as your target for using this tool. Do not, for instance, select all truck drivers in the facility. Select a group from that classification, such as shipping, receiving, tugger drivers, material handlers, or material deliveries. The size of your cross-functional team will be determined by the amount of people you are going to study in a specific group. For instance, if you are going to study three people, you will need a minimum of six on your cross-functional team and, if you are going to study six people, you will need a minimum of 12 people on your cross-functional team. You

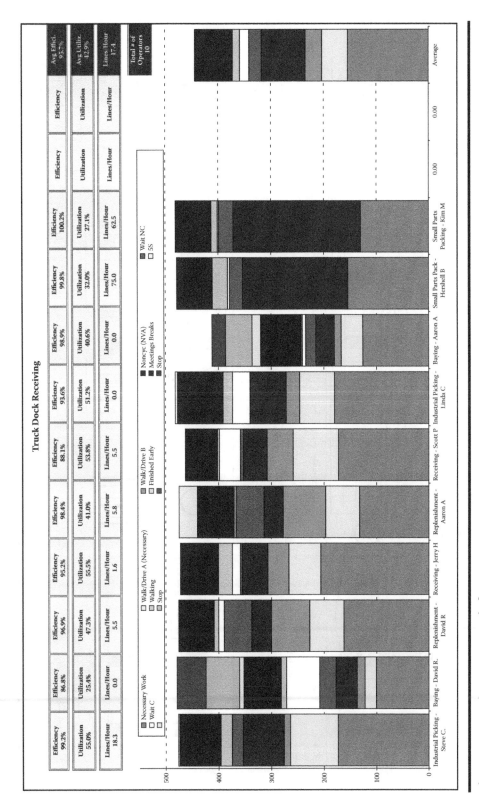

Figure 1.7 Group yamazumi chart.

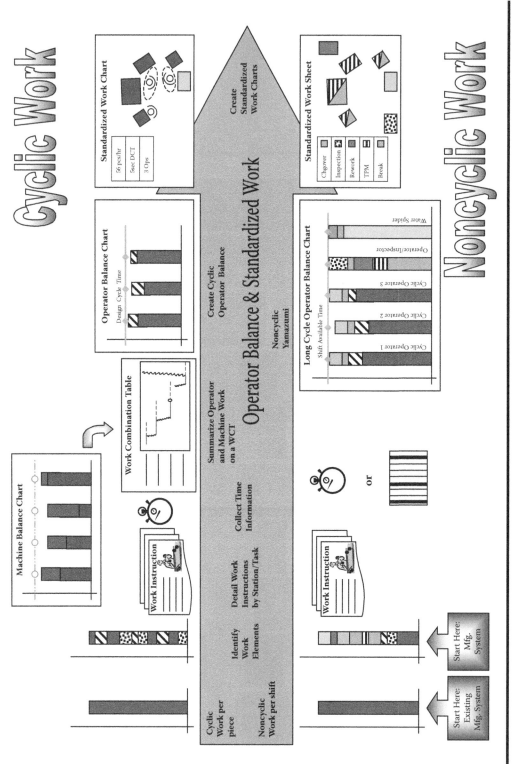

Figure 1.8 Cyclic versus noncyclic process flow chart.

can always have more, but you need a minimum of two team members per person being studied. You also will need to use a common sense approach in the selection of the team members. It must be cross functional with people from the group, their customers, team leaders, supervisors, etc. I also like to throw in what I call a ringer in every cross-functional team; someone who has nothing to do with the targeted job. It may be someone from HR, finance, or purchasing organizations. These people are helpful because they will always ask, "Why do you do that?" They bring a fresh set of eyes to the team process and always make a contribution before the end of the exercise.

Third Prerequisite

You must be able to identify what the current actual work practices are and understand the work practices for each position to classify these as "nonvalue add but necessary" and "nonvalue add" work and/or practices. An additional aid in clarifying what would be considered nonvalue add but necessary and nonvalue add is to determine exactly what work they are being paid to accomplish. This is where team members from management and the group that are the focus of the exercise play an important role. They can tell you exactly what they do all day and the various tasks they perform. I also guarantee that on more than one occasion they have said to themselves, "There has to be a better way to do this."

Fourth Prerequisite

You will need to make external preparations for your workshop and one full shift to observe the job being reviewed. You also will need to secure a room for your workshop for anywhere from three to five days, depending on how many people you are going to actually review. The more people under review, the longer it will take the team to collate the data once the observation is complete. The room needs to be large enough to accommodate the size of your team. Allow a large work table for the team and space for extras like easels with chart pads and portable dry erase boards, plus walls on which to hang chart paper. Figure 1.9 provides a checklist of suggested materials for your workshop.

We shall examine some noncyclical thought-provoking work practices in the upcoming chapters as we move forward in our exercise and data collection. Also, we will look at our decision making on where, when, and for what areas we need to create standard work documents, and determine

Workshop Checklist

Item					Completed
Conference/Training Room					
Laptop computer					
Laptop projector					
Pencils for all team members					
Lined writing pads for all team members					
Folders to hold papers					
2 Chart pads on easels					
	(3M sticky pad type if possible, Scotch masking tape or pins if not possible)				
One dry erase board					
Dry erase and permanent markers.					
	Black, blue, green, and red				
Stop watches with lap capablities for each team member					
	Save some time prior to the observation process to allow Team members to practice the use of their stop watch				
Clipboards for team members to use during data collection					
2 to 4 calculators depending on team size					
Top Leadership Kickoff stating support, goals & objectives					
Team introductions					

Figure 1.9 Workshop checklist.

what we consider to be a nonvalue add but necessary activity or nonvalue add activity for the particular noncyclic activity.

This determination will be a specific step-by-step exercise to ensure that all participants understand the purpose, how to use the tools, how to collect the data, how to read and interpret the data collected, complete the Do–Check–Act portion of the PDCA cycle, and, most importantly, buy into the changes that will be made so an ownership is established for sustaining the changes and making continuous improvements. This is how you develop a Lean culture in the workplace. By following this methodical group approach, you will ease concerns or fears of change, eliminate the "Flavor of Month" mentality, and begin the associate empowerment and input process.

Chapter 2

Identifying the Current Work Categories

This chapter covers how to identify the categories of work within a specific job title. It is imperative to include representation from the job being reviewed, across shifts if necessary, on your cross-functional team, to ensure optimum information exchange. The purpose of this chapter is to capture the tasks that are performed by the work group, even if these tasks are not viewed as being within the normal scope of work expectation. This step will set the foundation for your data collection categories that will be used in the master yamazumi later, so it is critical that the tasks are identified correctly. (See Chapter 1: Work Combination Table section for definition of yamazumi.)

Getting Started

The first stage of the Plan-Do-Check-Act (PDCA) cycle that was defined in Chapter 1 is to identify the current work categories; this would be considered part of the Plan stage. It is very important that we understand what the current Standard Work categories actually are, not what we think they are or what the work instruction or engineering routing says they are. This is why cross-functional groups are so important when launching a noncyclical Standard Work exercise. It is required that you have a minimum of one representative present from the noncyclical process under study; more than one is better because the information will be more representative of an entire group. Further, if you have multiple shifts, you should have at least one representative from each shift. After all, you are trying to determine the current state and, the more accurate the information you have, the better results you will receive from your data collection, which drives the decision-making process of improvements.

Once you have a cross-functional group organized, you need to announce the purpose of this first stage; simply to identify what work categories are actually being performed by the noncyclical group being studied. At this stage, it is important not to be judgmental about the work categories identified or make a decision regarding whether the group represented should or should not be doing these tasks. This discussion will take place later in the exercise.

As with any exercise designated to make improvements, you will need data collection and data collection documents. Let's take a look at our first data collection document: the Job Category Template and Document Type (Figure 2.1).

The purpose of this document is two-fold. The primary purpose is to identify the work categories that occur while performing the selected job. This does not include walk, wait, or idle times at this point. The information relating to the work categories sets the foundation for your data collection. Again, I cannot stress enough the importance of this step. If

Job Category Template & Document Type

Job Title:										Team Members:
Task Description										
Standard Work Tool Required										
Standard Work Chart										
Time Measurement										
Standard Work Sheet										
Work Combination Table										
Work Balance Chart										
Noncyclical Yamazumi Chart										

Figure 2.1   Job category template and document type.

not done correctly, you will not have a solid foundation on which to build your data collection, therefore, you will have potentially collected not only inaccurate data, but data that will not provide information for making continuous improvements. Additionally, when the team goes to collect live data, there will be many disruptions to that process because the team members will start to observe work categories that have not been previously identified.

The secondary purpose of this document is to decide what Standard Work documents should be utilized for each identified work category to perpetuate continuous improvement.

Use the following steps to complete this document:

- Start by entering the name of the specific job and the number of operators that will be reviewed next to the section labeled **Job Title**.
- Next, begin a discussion or brainstorming session with the cross-functional team members. Start making a list of work categories on a chart pad or dry erase board. When the team believes it is complete, fill in the document labeled **Task Description**.
- The final step is to have the team identify what Standard Work documents should be applied to the various tasks listed in the columns and simply place a Yes or No in the rows across under the label **Standard Work Tool Required**. The noncyclical yamazumi replaces the standard Work Balance chart normally used for cyclical processes. Now is a good time for the team to review any relevant Standard Work documents or routings that already exist. The team should compare these documents to verify if they are still applicable to the job being reviewed. If they are, keep them; if not, place this as a "homework" assignment on your chart pad to update these documents as soon as possible for each section identified with a Yes. If there are no documents to review, then the homework assignment is to create the proper documents for each section marked with a Yes.

Completed Examples for Your Reference

Let's look at several examples of completed Job Category Template and Document Type charts. I have always appreciated the fact that when I am being trained on a new (or any document) process to have a, or some completed examples, keeping in mind these are examples for the purpose of enhancing your understanding. I hope you will find these examples helpful as you use this form.

Multiple Operators, One Machine

In our first example, there are three manufacturing operators assigned to oversee and maintain one machine, which is almost as long as a football field (Figure 2.2). The machine runs in "automatic mode" until an alarm would sound, the raw or work-in-process (WIP) material broke somewhere in the process, or a changeover was required. The tasks identified overlapped with each operator. The machine had been identified as having poor run time, hence, the request for the workshop. All three positions were reviewed and one operator from each position per shift was a member of the cross-functional team for this particular workshop.

Order Pickers Distribution Center

In this example, the activities of five order pickers in a distribution center will be tracked (Figure 2.3). The distribution center was experiencing a large amount of back orders and poor customer delivery performance. World-class order picking for a distribution center is 109 lines per hour and the distribution center's current state was 48. An inventory analysis showed that the material needed to fill customer orders was onsite, therefore, the location requested a noncyclical workshop to determine why there was a delay in filling orders in a timely fashion. All five order pickers were part of the cross-functional team. Since implementing noncylical Standard Work, they have increased to 102 lines per hour.

One Operator, Multiple Machines

Our next example describes the circumstance and activities of a job where one operator handles multiple machines throughout the work shift (Figure 2.4). The department was experiencing a large amount of machine lost time as well as idle time. The plant had a significant quantity of customer back orders and a poor customer delivery record. The location requested a workshop to identify causes for less than world class performance metrics. Four operators and two representatives from the customer were part of the cross-functional team.

Job Category Template & Document Type

Team Members:

Job Title:	Yarn Treatment Machine—3 Operators									
Task Description	Change-overs	Alarms	Load Creel / Tie Ends	Autonomous Maintenance	Unplanned Downtime	Wrapping & Stacking	Weighing	Quality Checks	Stocking	5 S
Standard Work Tool Required										
Standard Work Chart	No	No	No	No	No	No	No	No	No	No
Time Measurement	Yes	No	No	No	No	Yes	Yes	No	No	No
Standard Work Sheet	Yes	Yes	Yes	No	No	Yes	Yes	Yes	Yes	Yes
Work Combination Table	Yes	Yes	Yes	No	No	Yes	Yes	Yes	Yes	No
Work Instructions	Yes	Yes	Yes	Yes	No	Yes	Yes	Yes	Yes	Yes
Noncyclical Yamazumi Chart	Yes	Yes	Yes	Yes	Yes	Yes	Yes	Yes	Yes	Yes

Figure 2.2   Job category yarn machine example.

Job Category Template & Document Type

Job Title:	Picker—Distribution Center (5 Operators)									Team Members:
Task Description	Picking Order	Packing Order	Drive With Material	Drive With No Material	Search for Boxes	Change Battery	Drive to Secondary Location	Stocking	5 S	
	Standard Work Tool Required									
Standard Work Chart	No	No	No	No	No	No	No	No	No	
Time Measurement	Yes	Yes	Yes	Yes	Yes	Yes	Yes	Yes	Yes	
Standard Work Sheet	Yes	Yes	Yes	No	No	Yes	Yes	Yes	Yes	
Work Combination Table	No	No	No	No	No	No	No	No	No	
Work Instructions	Yes	No	Yes	No	No	Yes	Yes	Yes	Yes	
Noncyclical Yamazumi Chart	Yes	Yes	Yes	Yes	Yes	Yes	Yes	Yes	Yes	

Figure 2.3 Job category order picker example.

Job Category Template & Document Type

Job Title:	Screw Machine Department Team Leaders											Team Members:
Task Description	Run Machines	Problem Solve	Create WO Maint	Audit EP	Audit 5S	Chasing Tools	Chasing Spare Parts	Update Comm Boards	C/Os	Meetings	Co-ordinate Training	Manpower Adj.
						Standard Work Tool Required						
Standard Work Chart	No	No	No	No	No	No	No	No	No	No	No	No
Time Measurement	Yes	Yes	Yes	Yes	Yes	Yes	Yes	Yes	Yes	Yes	Yes	Yes
Standard Work Sheet	Yes	Yes	Yes	Yes	Yes	Yes	Yes	Yes	Yes	No	No	No
Work Combination Table	No	No	No	No	No	No	No	No	No	No	No	No
Work Instructions	Yes	No	Yes	Yes	Yes	Yes	Yes	Yes	Yes	No	Yes	Yes
Noncyclical Yamazumi Chart	Yes	Yes	Yes	Yes	Yes	Yes	Yes	Yes	Yes	Yes	Yes	Yes

Figure 2.4   Job category screw machine example.

Maintenance Technicians

The maintenance technician is an important part of an important support group in the manufacturing world and usually the highest paid. Without their skills, though, a facility could experience huge productivity losses when an unplanned downtime event occurs. We expect them to react swiftly and efficiently to repair the equipment so that customer schedules can be met. Because of that expectation, we are obligated to provide them the with proper support to perform their work categories in meeting that expectation. In this particular example, representative technicians from the two different classifications were involved as well as three shifts that operate at this facility (Figure 2.5). The plant was in the process of transitioning from a total reactive style of maintenance to a more predictive and planned maintenance style. The maintenance organization knew there were opportunities to improve utilization time and, therefore, they requested a workshop.

Fork Truck Drivers

This is an example of a typical shipping and receiving area where multiple fork drivers work in an area that includes most of the facility (Figure 2.6). There were several issues at this facility including semi trucks not being unloaded in a timely fashion. This caused extra demurrage charges and primary and secondary stocking locations to be frequently empty even though the material was on the dock floor or in a truck waiting to be unloaded. This was a two-shift operation, so representatives of both shifts were part of the cross-functional team.

Conclusion

These are only a few of the typical noncyclical processes found in most manufacturing operations, but do not limit yourself to just these examples. The field of application is limited only by your thinking and vision. As mentioned in the Preface under the section Why Should I?, there is a mountain of untapped savings in nonmanufacturing areas as well. For example, think about the finance department where accounts payable,

Job Category Template & Document Type

Job Title:	Maintenance										Team Members:
Task Description	PM	Break-down	Work Order Documentation	Hoist Safety Inspections	Rebuild Off Line	Chasing Tools	Chasing Spare Parts	Job Site Cleanup	C/Os	Driving	
			Standard Work Tool Required								
Standard Work Chart	No	No	No	No	No	No	No	No	No	No	
Time Measurement	Yes	Yes	Yes	Yes	Yes	Yes	Yes	Yes	Yes	Yes	
Standard Work Sheet	Yes	Yes	Yes	Yes	Yes	Yes	Yes	Yes	Yes	No	
Work Combination Table	No	No	No	No	No	No	No	No	No	No	
Work Instructions	Yes	Yes	Yes	Yes	Yes	Yes	Yes	Yes	Yes	Yes	
Noncyclical Yamazumi Chart	Yes	Yes	Yes	Yes	Yes	Yes	Yes	Yes	Yes	Yes	

Figure 2.5 Job category maintenance example.

Job Category Template & Document Type

Job Title:	Fork Truck Drivers - Shipping & Receiving										Team Members:
Task Description	Load Semi Trailers from Staging	Unload Semi Trailers to Staging	Drive With Material	Drive With No Material	Search for Boxes	Change Battery	Drive to Primary Location	Stocking	5 S	Drive to Secondary Location	
	Standard Work Tool Required										
Standard Work Chart	No	No	No	No	No	No	No	No	No	No	
Time Measurement	Yes	Yes	Yes	Yes	Yes	Yes	Yes	Yes	Yes	Yes	
Standard Work Sheet	Yes	Yes	Yes	No	No	Yes	Yes	Yes	Yes	Yes	
Work Combination Table	No	No	No	No	No	No	No	No	No	No	
Work Instructions	Yes	Yes	Yes	No	No	Yes	Yes	Yes	Yes	Yes	
Noncyclical Yamazumi Chart	Yes	Yes	Yes	Yes	Yes	Yes	Yes	Yes	Yes	Yes	

Figure 2.6 Job category fork truck driver example.

receivables, and payroll are processed; the HR department where training and orientation programs are supplied; the engineering department where testing, product design, product development, and new program launches take place; the purchasing department where supplier contracts are issued. Do you begin to see the additional opportunities in your organization?

Chapter 3

Identifying Nonvalue Add but Necessary and Nonvalue Add

It is true that the only value adder of any process is the operator/associate who changes form, fit, or function of a product to meet customer requirements. All other personnel are considered nonvalue added (but necessary) to a given process or product. However, within all nonvalue-added work practices, there are work categories that we consider the "necessary" part of their supporting the main process or product.

Defining the Nonvalue Add but Necessary (NVA Necessary) and Nonvalue Add (NVA) for Noncyclic Work

Before we examine and explain the tool in the next step in the documentation process, let's talk about how we would define the nonvalue add but necessary (NVA Necessary) and nonvalue add (NVA) work categories for several typical noncyclical positions. You may even recognize these positions and habits at your facility. As a reminder here, any work that does not change the form, fit, or function for the customer is considered as NVA. I agree with that definition. What I am referencing here is the categories of work that are considered NVA Necessary to support an efficient operation on a daily basis.

Maintenance

An NVA Necessary practice for a maintenance technician would include repairing a piece of equipment that has experienced a breakdown, or a maintenance technician who performs a planned preventative maintenance work order (PM) to maintain equipment. An example of NVA practices for a maintenance technician would be walking to the MRO (maintenance repair and operations, or spare parts) crib for parts, searching for parts, or searching for required tools while a piece of equipment remains inoperable for manufacturing. Or, in the worst case scenario, waiting for something to happen so he/she can be called to a problem area. Generally speaking, a maintenance technician or skilled trades position is usually high paying due to the skill sets required to perform the tasks. Any improvements made in this area will make a significant contribution to improving flow and we know that, when we improve flow, we improve the impact to our costs.

Order Picker

An NVA Necessary practice for an order picker in a distribution center would be the actual picking of a customer order or driving to deliver that order to the packing area. Notice here that the order picker in these scenarios is actually handling material, not driving or walking empty handed. Several NVA examples for an order picker would be searching for his/her vehicle, waiting for material to be placed in the primary pick location, or gathering boxes in which to place customer-ordered parts or driving or walking empty handed.

Order Packer

An NVA Necessary practice for a packer in a distribution center would be the actual packing for shipment of a customer order that has been delivered by some method, such as the order picker or a conveyor. Several examples of NVA practices for an order packer would be waiting for a customer order to be delivered, searching for a customer order that needs to be shipped, or searching for supplies to pack a customer order for shipment or waiting for materials to be packed for a customer order.

Bin Stockers

An NVA Necessary practice for a stocker in a distribution center would include driving with material on a vehicle that is be delivered to the primary stock location or unloading shipments received for stocking. Several examples of NVA practices for a stocker would be driving with the loaded vehicle to move material from one location to another simply to free up floor space, unloading unscheduled shipments received, or driving an empty vehicle.

Material Handlers or Water Spider for Manufacturing

An NVA Necessary practice example for a material handler or water spider would be delivering the right material to the right location at the right time. Several examples of NVA practices for a material handler or water spider includes driving around without parts, returnable dunnage containers, or recycle materials scrap or trash; delivering the wrong parts to manufacturing; poorly timed delivery routes, or searching for material to deliver to manufacturing.

Qualifying Work Categories

The next step is qualifying what work categories would be considered NVA Necessary and NVA. Using the previously suggested points should help put one on the right path. This document is called the Work Summary Sheet (Figure 3.1). This document is a multipurpose tool that defines the work categories into color-coded NVA Necessary and NVA groupings (see the color figure on the CD that accompanies the book). We also will use this document to place our data regarding total times collected during our observation exercise. Remember this important point: This is not an exercise to compare one operator against another. It is an exercise to gather data to see the current state and use that as a baseline to make continuous improvements. Also, it is to bring all the operators to a level playing field when performing their work. It is essential that the cross-functional team members and the group being studied understand this. We are reviewing processes, not people.

To complete this document, follow these steps:

1. Start by entering the line or process name and the date in the appropriate areas.
2. Next determine the **Gross Available** minutes, minus any minutes for scheduled breaks, lunch scheduled meeting times, and any nonwork time planned for the operator. This will determine the net available minutes per shift. For example:

Noncyclic Work Summary Sheet

Line Name/Process Name: _____

Date: _____

Gross Available Minutes for Shift: _____

Minutes at Meetings, Breaks & Meals: _____

Net Available Minutes for Shift: _____

Work Element ID	Work Description	Min / Task	# Times Repeated	Cumulative Time (min)
Work NVA Necessary				
Work NVA Necessary				
Work NVA				
NON CYCLIC NVA				
WAIT NO CHOICE NVA				
WAIT CHOICE NVA	Leaving early or coming back late, stopped to talk to coworkers, etc.			
WALK NVA	Any walking			
			Total	0

Figure 3.1 Work summary sheet (see the color figure on the CD that accompanies the book).

Shift is 8 hours = 480 gross available minutes minus 2- to 15-minute breaks a shift, a 30-minute lunch, and a daily 10-minute shift start meeting = 410 net available minutes.

480 minutes minus 30 minutes minus 30 minutes minus 10 minutes = 410 available minutes.

3. Place this information in the appropriate areas on the Work Summary Sheet.

Look at the noncyclic color-coded work elements chart in Figure 3.2. You will notice that there are two different color codes for NVA Necessary work categories and five different color codes for NVA actions. The reasons for these categories are:

Color Coded Work Elements

Necessary Work: (*NVA*) Activity that is
essential to the support of the production
process.

Non-Necessary Work: (*NonVA*) Any activity that
is nonessential to the production process.

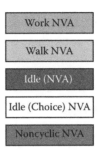

**Figure 3.2   Noncyclic color-coded work elements (see the color figure on the CD
that accompanies the book).**

- The green NVA Necessary is for the primary duties of the operators. These are work categories that, based on their type of work, are the functions that operators should be performing for the largest percentage of their work day.
- The yellow NVA Necessary is for other work categories that need to be performed, but we hope to find fewer of their net available minutes showing up in this section.
- The orange NVA color code is the list work that is being performed outside of either their green VA or yellow VA.
- The grey NVA color code is used to capture any walking.
- The red NVA color code is to capture any waiting time where the operator has no control over the condition.
- The white NVA color code is to capture any waiting or delays where the operator has control over the condition.
- The blue NVA color code is mainly to capture time spent looking or searching for items necessary to complete their work categories.

Completed Examples for Reference

In this section, we look at several examples of completed Work Summary Sheets, which are part of the data collection documentation. It is important to retain and track this documentation throughout the process for reference later in case questions arise, to address any challenges about how conclusions were made, and remove emotion from the process and keep it focused on data-driven need for change.

Yarn Winding Operator

After reviewing the example in Figure 3.3, you may have questions regarding the category definitions, so let's discuss some areas. Notice that the

Noncyclic Work Summary Sheet

Line Name/Process Name: Operator Yarn Winding

Date: June 15, 2007

Gross Available Minutes for Shift: __480 min__

Minutes at Meetings, Breaks & Meals: __40 min__

Net Available Minutes for Shift: __440 min__

Work Element ID	Work Description	Min / Task	# Times Repeated	Cumulative Time (min)
Work NVA Necessary	Change Over, Package, Creels, Weigh, Work Order, 5S			
Work NVA Necessary	Material on Forks			
Drive NVA	No Material on Forks			
NON CYCLIC NVA	Alarm Response, Quality Checks			
WAIT NO CHOICE NVA	Planned Maintenance, Unplanned Downtime, No Material			
Idle Choice NVA	Leaving early or coming back late, stopped to talk to co-workers, etc.			
Walk	Walk			
Note: Tag Relief Accounts for Minutes Beyond net Avilable Minutes for Shift.			Total	0

Figure 3.3   Yarn winding work summary NVA (see the color figure on the CD that accompanies the book).

only work considered as green NVA Necessary in this facility is the actual changeover, weighing and packaging a customer order, completing a work order, and 5S tasks. (5S tasks are those that fall under the category of Sift-Sort-Sweep-Standardize and Sustaining a Work Area for housekeeping, safety, and visual factory to determine normal from abnormal conditions.) This is the primary work of this operator. They also recognize that this particular operator has to do some fork truck driving (yellow and orange) throughout the shift, but, on this particular snapshot of time, he or she did none. The blue section is part of the operator's responsibility, but his/her goal is to eliminate the causes for alarms and quality checks; this is why they marked it as NVA. The red, wait no choice, has been defined as planned maintenance because the machine runs in auto for 24 hours a day, thus the machine must shut down for any PMs. The remaining categories should not require further explanation.

Order Picking

The operator classification is that of an order picker and, if customers orders aren't picked, then they cannot be shipped (Figure 3.4). The facility also has a group classification of packers. Once we see the entire group activity, we will be able to get into the problem-solving stage.

Screw Machine Operator

This example is of a team leader whose role is to support the operators and not specifically to run the machines (Figure 3.5). Additionally, roles include preparing the external elements for upcoming changeovers. It is also expected that the team leader replace operators in case of high absenteeism.

Maintenance Technician

Figure 3.6 shows a typical breakdown of the roles of a maintenance technician who is assigned to support a manufacturing process and be on call to react to any unplanned downtime. Now, if you are studying a group of maintenance technicians in a tool room, for example, your NVA Necessary and NVA choices will probably look a bit different. As a reminder, your NVA Necessary should capture what the main work expectation is, i.e., what these people are being paid to do.

Noncyclic Work Summary Sheet

Line Name/Process Name: Order Picker Operator B

Date: February 25, 2009

Gross Available Minutes for Shift: __480 min__

Minutes at Meetings, Breaks & Meals: __47 min__

Net Available Minutes for Shift: __433 min__

Work Element ID	Work Description	Min / Task	# Times Repeated	Cumulative Time (min)
Work NVA Necessary	Picking: Remove product from shelf & place in box on vehicle			
Work NVA Necessary	Driving activity from pick location to pick location prior and delivery to pack point			
Work NVA	All other driving activity. Packing			
NON CYCLIC NVA	Building boxes, change battery in scanner, change truck batttery			
WAIT NO CHOICE NVA	Waiting on stock in primary, out of stock, waiting on vehicles, waiting on wrapper, waiting for people, verbal instructions.			
Wait CHOICE NVA	Restroom, smoke, emergency phone calls, talking, helping others, finished early			
WALK NVA	Any walking			
			Total	0

Figure 3.4 Order picker work summary NVA (see the color figure on the CD that accompanies the book).

Fork Truck Driver Shipping/Receiving

Fork truck driving positions are an opportunity that should not be overlooked; there are huge efficiency savings in most of these assignments (Figure 3.7). This is not because truck drivers don't want to work, it is because of the obstacles that prevent them from performing their duties in a manner that supports flow. Any time a fork truck driver is moving with an empty fork, it is essentially waste.

Noncyclic Work Summary Sheet

Line Name/Process Name: <u>Team Leader Screw Machine</u>

Date: <u>May 7, 2008</u>

Gross Available Minutes for Shift: <u>480 min</u>

Minutes at Meetings, Breaks & Meals: <u>40 min</u>

Net Available Minutes for Shift: <u>440 min</u>

Work Element ID	Work Description	Min / Task	# of Times Repeated	Cumulative Time (min)
Work NVA Necessary	Updating communication boards and huddle meetings, problem solving, Error Proofing audits, 5S audits, preparing external actions for change overs			
Work NVA Necessary	Coordinate training, manpower adjustments			
NVA	Replace missing operators and run the machines			
NON CYCLIC NVA	Chasing parts and tools, work related phone conversations to order tooling			
WAIT NO CHOICE NVA	Waiting for machines, trucks, computers, information, maintenance or tools			
WAIT CHOICE NVA	Leaving early or coming back late, stopped to talk to coworkers, etc.			
WALK NVA	Any walking			
			Total	0

Figure 3.5　Screw machine operator NVA (see the color figure on the CD that accompanies the book).

Completed Forms

Once you have completed your form(s), you now have the basis for your data collection sheets that need to be prepared prior to going to the floor, shadowing individuals, and collecting the actual times for the various categories. We will be using these forms again once that process is completed to input the total amount of time for the various categories and

Noncyclic Work Summary Sheet

Line Name/Process Name: Maintenance-Working in Manufacturing Area

Date: September 20, 2008

Gross Available Minutes for Shift: ___480 min___

Minutes at 5s, Breaks & Meals: ___50 min___

Net Available Minutes for Shift: ___430 min___

Work Element ID	Work Description	Min / Task	# Of Times Repeated	Cumulative Time (min)
Work NVA Necessary	PM activity (prep, job, components). Removing guards, panels, housekeeping, job site clean up, work order documentation, checking on computer terminal required to do job, other required documentation. Safety inspection of hoists.			
Work NVA Necessary Berakdown	Anything related to unplanned activity, rebuild equipment off line while machine is running, fixing machine during unplanned downtime, unplanned changeovers, clean up due to breakdowns.			
NVA Drive	Driving equipment with forks or cargo bay empty as in going to get parts or to a job like JLG lift. Driving to hoist inspection sites.			
NVA NON CYCLIC	Battery changes, maintaining maintenance equipment, clearing an area to work on equipment, having to make a part-jig or tool to complete a job.			
NVA Idle CHOICE	Leaving early. Stopped to talk to co-workers, restroom, smoke, emergency or other phone calls.			
NVA Wait NC	Waiting for Operator, parts or instructions, aisle blocked by parts or equipment, blocked path.			
NVA walk	Walking looking for parts, walking to a job, walking or searching for/to a piece of power equipment (forklift, etc.).			
			Total	0

Figure 3.6 Maintenance NVA (see the color figure on the CD that accompanies the book).

Noncyclic Work Summary Sheet

Line Name/Process Name: <u>Fork Truck Driver Shipping-Receiving</u>

Date: <u>December 5, 2008</u>

Gross Available Minutes for Shift: <u>480 min</u>

Minutes at Meetings, Breaks & Meals: <u>50 min</u>

Net Available Minutes for Shift: <u>430 min</u>

Work Element ID	Work Description	Min / Task	# Times Repeated	Cumulative Time (min)
Work NVA Necessary	Receiving containers/cartons, put away stock, load semi trailers, 5S			
Work NVA Necessary Drive	Drive to put away stock			
NVA Drive	Driving empty, to remove an empty pallet, breakdown pallet to unwrap, changing battery			
NONCYCLIC NVA	Set up cart, removing cartons and trash, pre-sorting to put away stock, fixing disturbed parts, walk to find ladder or pallet			
WAIT No CHOICE NVA	Waiting on pallet drop downs, waiting on vehicles, waiting for people, verbal instructions, check report, work related phone call			
WAIT CHOICE NVA	Restroom, smoke, personal phone calls, talking with others, leave early or return late			
WALK NVA	Any walking			
			Total	0

Figure 3.7 Fork truck driver shipping/receiving NVA (see the color figure on the CD that accompanies the book).

transferring those numbers to our noncyclic yamazumi. It is critical that during this process that the NVA Necessary and NVA categories are clearly identified and agreed to up front. The only reason I bring this up is that I have experienced businesses in the past that, when the data collection shows them something they were not happy with or the data tell them something they don't want to hear, they claimed that the NVA Necessary and NVA categories were not identified properly. They basically go back and change the categories so it will skew the data to show and tell them what they want to hear, not the true current state. By doing this, they are able to justify doing nothing. Do not fall into this trap; remember, when we embark on a Lean journey, we frequently receive information that does not look good or is unpleasant to hear. But, this is not about someone or some department looking bad. It is about creating continuous improvement to our process so we can be competitive in a global market even to keep the plant open. We must get beyond the point where our egos or pride won't allow us to make good business decisions based on accurate and truthful data.

Chapter 4

Collecting and Collating Data

In this chapter, we come to the most critical portion of the workshop exercise: collecting and collating the data. We will discuss two methods of collecting the data: (1) the "old fashioned" way, by using a stopwatch, and (2) a newer method, using barcodes and a barcode reader device.

I describe using a stopwatch as the "old fashioned" method because this is the tool most commonly used to teach people to monitor a process or gather accurate data. Even in today's world, though, using stopwatches on the shop floor has a tendency to make operators nervous and, thus, they sometimes complete job elements differently. Also, I have encountered businesses that have a union partner where using a stopwatch on the manufacturing floor is not permitted by contractual language.

Therefore, an alternative method was devised that accurately collects data, but does not create the tension sometimes caused by the use of a stopwatch to verify a process. The barcode method is less intrusive and is usually acceptable in most conditions. Of course, if there is contractual language regarding this type of activity, advance discussion is required with the local union representation to clarify any misgivings before proceeding with either the stopwatch or barcode method.

Collecting the Data

This process takes place in the manufacturing environment with old-fashioned footwork and observations.

Remember what was said in Chapter 1, section Second Prerequisite: team members should be split into groups of two (minimum) to three people and be assigned to observe one of the people in the group who is being studied, one person timing, and another documenting. Once you have your groupings, you should introduce each team member to the person he/she will be shadowing the next day. Allow them some breakout time to introduce themselves, to plan where to meet prior to the beginning of the shift, understand when this person takes breaks and lunch times, etc. The groups need to understand the working hours of the individual they will shadow as they must work the same hours. You may have groups working first, second, or third shifts, depending on the facility.

A data collection document will need to be prepared showing all the nonvalue add but necessary (NVA Necessary) and nonvalue add (NVA) work categories for your cross-functional team members to use on the shop floor. This is done based on what was just completed on the noncyclical Work Summary sheets (see Figure 3.1). It is very important to list all of the categories from that document onto a time data collection form that clearly identifies the various sections, even if the subcategories have been identified to fit into the same NVA Necessary or NVA. This information detail will likely be useful when you start to problem-solve with regard to reducing the NVA and increasing the NVA Necessary. These documents must be user friendly, i.e., categories must be clear and easy to understand and to find on the collection document. You do not want people searching for categories or the location of a document when they are on the manufacturing floor.

Now is a great time for another Plan-Do-Check-Act (PDCA) cycle. One important point to impress upon those collecting time measurements is the *starting* and *stopping point* of each NVA Necessary or NVA action documented. On the previous day, allow at least three hours for each miniteam to go out to the floor and practice, practice, practice. When the team members return, have an open discussion about what transpired on the floor and make any corrections to the data collection document that may be necessary. Ensure that all groups understand the plan for data collection the next day, who they are going to shadow, where they will meet, and any required safety equipment the team members need for the manufacturing environment. Remember the old adage: A plan does not guarantee success, but not having a plan guarantees failure.

Using a Stopwatch to Collect Data

The stopwatch is a tried and true (and historic) device used for data collection. For our purposes here, stopwatches with lapping capabilities will need to be provided so the people collecting the data will not lose increments of time in between the starting and stopping of different sequences of actions. Sharpened pencils and clipboards also will need to be provided.

Figure 4.1 shows a simplified example of one noncyclic job (material deliveries) to give an idea of what I am referencing; however, one could also create his own style of document to better match his specific needs.

The layout of the data collection form is not as important as having the team members develop a document that will work for them. After all, they are the ones who will be using it. Once you have a data collection document developed, provide a brief training session for the team members to learn how to use a stopwatch and how to use it in a manner for their collection purpose.

Using Barcodes to Collect Data

Your facility may have hand-held barcode scanners or PDAs (personal digital assistants) that have barcode scanning capabilities available. There also is a device on the market called Opticon that works very well in any environment. It is small, cheap, and easy to use as well as more accurate, efficient, and error proof. This methodology also allows you to have a one-on-one ratio for collecting data. When creating your data collection sheet, simply duplicate the categories and have your IT department "translate" the category into a barcode that will be scanned each time the activity takes place. This barcode then will be placed on the sheet next to each category. Once a barcode is scanned, all time is attributed to that category until another barcode is scanned. The barcode scanning device will track the time as well. If you happen to scan the wrong barcode, not to worry. Simply scan the correct one quickly. Remember, there should be time allotted to practice the barcode method just as you would if you were using stopwatches.

My experiences to date, based on the feedback of the users, is that they much prefer the barcode scanning method. This is true also of the people

Category	Time Collection of Each Occurrence
Drive with Forks/Cart Empty	
Drive with Material	
Deliver Material to Device	
Wait (No Choice)	
Wait (Choice)	
Walk	
Process Paper Work	
Idle	
Finish Early	

Figure 4.1 Stopwatch data collection example.

operating the processes being reviewed as well as management and unions alike. See Figure 4.2 for a simplified example of a data collection sheet using barcodes.

At the end of the observation, the barcoding device data collected are downloaded into the master yamazumi where the information will place the data into the proper categories and generate the yamazumi. Also, the data will automatically drop into the charts and perform calculations. No manual data entry is required. Once you have a data collection document developed, provide a brief training session for the team members to learn how to use a scanning device and, as in the stopwatch method, how to use it in a proper manner for their purpose.

If you use stopwatches to collect the data, you will need to summarize the data on a Work Summary Sheet (note the completed examples in Figure 4.3 through Figure 4.7). The Work Summary Sheet will help with the manual data entry into the master yamazumi individual section, so the main group charts will populate.

If you have used barcode scan sheets and barcode scanning devices, this step is not necessary.

Collating the Stopwatch Data

The miniteams will have spent an entire day collecting data either by stopwatch or barcoded scanner. Now, you will need to verify, again the PDCA cycle, for accuracy.

If stopwatches were used, the miniteams will need to add up all the times from the various NVA Necessary and NVA categories using a calculator. Once this is completed, go back to the noncyclical Work Summary Sheet and populate the times for each appropriate category.

If barcode scanners were used, download the data into the master yamazumi file. Note the amount of time given to each NVA Necessary and NVA and transpose this information onto the noncyclic Work Summary Sheet.

For both methods, the total at the bottom of the sheet should match the "Net Available Minutes" you input at the top of the sheet. If they do not match, the team will need to review their data for errors. If they do match up, and even if they don't, be sure to congratulate the team on a job well done; this was a difficult task.

Process ID		12-Jan-09	Operator A	B	Barcode	Process ID		Process ID
Barcode Here	1	Drive VA	Pick up pallet, Label Pallet, Stretch Wrap, Put in staging area		Barcode Here	8	Finished Early	Finished Early
Barcode Here	2	Drive VA	Driving from drop off area, driving to staging area		Barcode Here	9	Meetings - Breaks	Minutes for Meetings, Breaks
Barcode Here	3	Drive NVA	Driving Empty		Barcode Here	10	Out of material	Material not in stock anywhere on grounds
Barcode Here	4	Noncyc	Changing batteries, making boxes, etc.		Barcode Here	11	No Stock	Out of Stock, no material on facitliy
Barcode Here	5	Idle No choice	Waiting on vehicles, waiting for people, verbal instruction, waiting for data to cycle, etc.		Barcode Here	12	Drive NVA	Primary location empty must drive to secondary location
Barcode Here	6	Idle Choice	Restroom, smoke, emergency phone calls, talking to others, idle		Barcode Here	13	Wait	Waiting for material to be stocked in bin
Barcode Here	7	Walking	Any walking		Barcode Here	14	Not listed	Not listed on sheet will document title by hand

Figure 4.2 Scan sheet example.

Noncyclic Work Summary Sheet

Line Name/Process Name: <u>Operator Yarn Winding</u>

Date: <u>June 15, 2007</u>

Gross Available Minutes for Shift: _____ **480 min** _____

Minutes at Meetings, Breaks & Meals: _____ **40 min** _____

Net Available Minutes for Shift: _____ **440 min** _____

Work Element ID	Work Description	Min / Task	# Times Repeated	Cumulative Time (min)
Work NVA Necessary	Change Over, Package, Creels, Weigh, Work Order, 5S	Varies	5	288
Drive NVA Necessary	Material on Forks	Varies	0	0
Drive NVA	No Material on Forks	Varies	0	0
Noncyclic NVA	Alarm Response, Quality Checks	Varies	11	60
Wait No Choice NVA	Planned Maintenance, Unplanned Downtime, No Material	Varies	7	46
Idle Choice NVA	Leaving Early or Coming Back Late, Stopped to Talk to Coworkers, etc.	Varies	3	34
Walk	Walk	Varies	8	39
Note: Tag Relief Accounts for Minutes beyond Net Available Minutes for Shift			Total	466

Figure 4.3   Yarn winding noncyclic work summary with data.

Completed Examples

Let's look at several examples we have been using throughout the book that have been populated with data that was collected. We will be using these to populate information into the individual yamazumi for each person that has been observed for this exercise (Figure 4.3 through Figure 4.7).

Noncyclic Work Summary Sheet

Line Name/Process Name: <u>Order Picker Operator B</u>

Date: February 25, 2009

Gross Available Minutes for Shift: ____480 min____

Minutes at Meetings, Breaks & Meals: ____47 min____

Net Available Minutes for Shift: ____433 min____

Work Element ID	Work Description	Min / Task	# Times Repeated	Cumulative Time (min)
Work NVA Necessary	Picking: Remove product from shelf & place in box on vehicle	Varies	6	28
Work NVA Necessary	Driving activity from pick location to pick location prior and delivery to pack point	Varies	4	38
Work NVA	All other driving activity. Packing	Varies	32	228
NONCYCLIC NVA	Building boxes, change battery in scanner, change truck batttery	Varies	3	94
WAIT NO CHOICE NVA	Waiting on stock in primary, out of stock, waiting on vehicles, waiting on wrapper, waiting for people, verbal instructions	Varies	1	7
WAIT CHOICE NVA	Restroom, smoke, emergency phone calls, talking, helping others, finished early	Varies	3	15
WALK NVA	Any walking	Varies	8	23
			Total	433

Figure 4.4   Order picker noncyclic work summary with data.

During the review of the noncyclic Work Summary sheets, this is a perfect opportunity to gather feedback of what the team members observed during the data collection process. When you are facilitating this exercise, make sure you take the time to capture the observations on a chart pad for further followup. Many times, the team members are in the perfect

Noncyclic Work Summary Sheet

Line Name/Process Name: Team Leader Screw Machine

Date: May 7, 2008

Gross Available Minutes for Shift: _____ 480 min _____

Minutes at Meetings, Breaks & Meals: _____ 40 min _____

Net Available Minutes for Shift: _____ 440 min _____

Work Element ID	Work Description	Min / Task	# Times Repeated	Cumulative Time (min)
Work NVA Necessary	Updating communication boards and huddle meetings, problem solving, Error Proofing audits, 5S audits, preparing external actions for change overs	Varies	1	25
Work NVA Necessary	Coordinate training, manpower adjustments	Varies	1	25
NVA	Repalce missing operators and run the machines	Varies		213
NONCYCLIC NVA	Chasing parts and tools, work related phone conversations to order tooling	Varies	6	59
WAIT NO CHOICE NVA	Waiting for machines, trucks, computers, information, maintenance or tools	Varies	3	11
WAIT CHOICE NVA	Leaving early or coming back late, stopped to talk to coworkers, etc.	Varies	6	48
WALK NVA	Any walking	Varies	8	59
			Total	440

Figure 4.5   Screw machine team leader noncyclic work summary with data.

position to observe some or most of the roadblocks that impact the processes being reviewed. Some examples I have encountered during theses sessions include:

■ Maintenance technicians having to walk to the opposite end of the plant to get spare parts for a repair.

Noncyclic Work Summary Sheet

Line Name/Process Name: Maintenance-Working in Manufacturing Area

Date: September 20, 2008

Gross Available Minutes for Shift: _____ 480 min _____

Minutes at 5S, Breaks & Meals: _____ 50 min _____

Net Available Minutes for Shift: _____ 430 min _____

Work Element ID	Work Description	Min / Task	# Times Repeated	Cumulative Time (min)
Work NVA Necessary	PM activity (prep, job, components). Removing guards, panels, housekeeping, job site clean up, work order documentation, checking on computer terminal required to do job, other required documentation. Safety inspection of hoists.	Varies	7	171
Work NVA Necessary Breakdown	Anything related to unplanned activity, rebuild equipment off line while machine is running, fixing machine during unplanned downtime, unplanned changeovers, clean up due to breakdowns.	Varies	2	45
NVA Drive	Driving equipment with forks or cargo bay empty as in going to get parts or to a job like JLG lift. Driving to hoist inspection sites.	Varies	0	0
NVA-Noncyclic	Battery changes, maintaining maintenance equipment, clearing an area to work on equipment, having to make a part-jig or tool to complete a job.	Varies	0	0
NVA Idle Choice	Leaving early. Stopped to talk to Coworkers, restroom, Smoke, emergency or other phone calls.	Varies	4	41
NVA Wait NC	Waiting for Operator, parts or instructions, aisle blocked by parts or equipment, blocked path.	Varies	3	60
NVA Walk	Walking looking for parts, walking to a job, walking or searching for/to a piece of power equipment (forklift, etc.).	Varies	9	113
			Total	430

Figure 4.6   Maintenance noncyclic work summary with data.

Noncyclic Work Summary Sheet

Line Name/Process Name: Fork Truck Driver Shipping-Receiving

Date: December 5, 2008

Gross Available Minutes for Shift: _____ 480 min _____

Minutes at Meetings, Breaks & Meals: _____ 50 min _____

Net Available Minutes for Shift: _____ 430 min _____

Work Element ID	Work Description	Min / Task	# Times Repeated	Cumulative Time (min)
Work NVA Necessary	Receiving containers/cartons, put away stock, load semi trailers, 5S	Varies	16	217
Work NVA Necessary Walk	Walk to put away stock	Varies	2	35
NVA WALK	Walking to get a ladder, to remove an empty pallet, breakdown pallet to unwrap	Varies	2	22
NONCYCLIC NVA	Set up cart, removing cartons and trash, pre-sorting to put away stock, fixing disturbed parts	Varies	5	90
WAIT NO CHOICE NVA	Waiting on pallet drop downs, waiting on vehicles, waiting for people, verbal instructions, check report, work related phone call	Varies	1	4
WAIT CHOICE NVA	Restroom, smoke, personal phone calls, talking with others, leave early or return late	Varies	1	9
WALK NVA	Any walking	Varies	6	52
			Total	430

Figure 4.7   Fork truck driver noncyclic work summary with data.

■ Maintenance technicians having to search through the MRO crib because the computer system says the spare part is in inventory, but no one can find it.
■ Maintenance technicians having to search through the MRO crib or even the plant because the spare part is not located where the computer system says it is supposed be or in the "normal" identified location.
■ Material handlers responding to unplanned calls for material delivery.

■ Fork truck drivers searching for parts in a central material storage location, but nothing is labeled or no storage locations are identified.
■ Fork truck drivers on the receiving dock who have their routines disrupted due to unplanned deliveries.

These are only a few of the more common roadblocks frequently identified. Do not waste this opportunity to capture these first-hand observation because they will be valuable to the team once the process problem-solving process starts.

Chapter 5

Populating the Individual and Group Yamazumi

The yamazumi is a great tool that not only provides data for decision making and strategy, but also provides a great visual display that drives the impact of the data to those viewing it.

The individual yamazumi clearly shows the time spent in the identified categories by one person of the process being reviewed. This is helpful, as you can see, by percentages; that is, the time spent on areas previously identified as nonvalue added but necessary (NVA Necessary) and nonvalue added (NVA).

The Group yamazumi shows how all of the individuals' activities compare to the same process being reviewed. The visual comparison here will help in not only identifying the waste to be eliminated, but also act as an aid in balancing the work between individuals in a group.

As a reminder, when reviewing the yamazumi data either as an Individual or Group chart, it is the process we are looking at correcting. Do not use this information for individual or group punitive action or you will lose credibility for this process and, more importantly, for yourself, and damage the Lean journey.

Individual Yamazumi

If you use stopwatches to collect your data, you need to verify that your data collection is accurate, then populate the individual yamazumi charts in your master yamazumi file.

Simply click on an Open Tab on the master yamazumi file. Enter the project name, person name or code, and the date studied. Notice that under the project name it is clearly identified as the Current State. Now simply place the time amounts from the noncyclical Work Summary sheet into column B next to the appropriate categories. As you input data, you will notice a stacked bar chart being created for you. Once you have completed inputting the data, this also will calculate efficiency and utilization for you. The final step is to change the name on the Open Tab to the process name or operator code you just identified on the individual yamazumi.

If you used barcode scanning to collect your data, the Individual charts and Group chart will automatically populate when you download the information. That is the beauty of using the barcode technology; it is a real improvement to doing all the manual data entry. Also, we know that with any manual data entry, it provides an opportunity for mistakes. However, in barcode scanning, you should review your numbers for accuracy and then fill in the information for each individual and change the Tab name just as directed above.

Just as in cyclical Standardized Work, this will establish our baseline for this noncyclic job category and provide information for continuous improvement. The target for the NVA Necessary of this noncyclic job should be the same as it is for a cyclic operator: 85% load. Generally, as stated earlier, what you will normally see is about a 20 to 29% load of NVA Necessary.

Completed Examples

Let's look at several examples we have been using throughout the book and highlight some points from each. Remember in these charts the colorcoded categories are not specific enough to explain the block of time in detail (see the color figures on the CD that accompanies the book). To get that you will have to refer back to your noncyclic Work Data Summary forms (see Figure 3.1).

You will note in Figure 5.1 that the Work and Drive, classified as NVA Necessary, makes up 35% of the workload. This is higher than one would normally expect to see in the first round of process review. However, there are three NVA categories that send a clear signal of requiring in-depth analyzing to make improvements: 29% of time is spent on noncyclic work, 10% of time is spent on Wait No Choice, 7% of time is spent on Idle Choice, and 11% of time is spent on Walk. These percentages of NVA equal 57% of the total time. These categories and percentages are fairly typical on the first

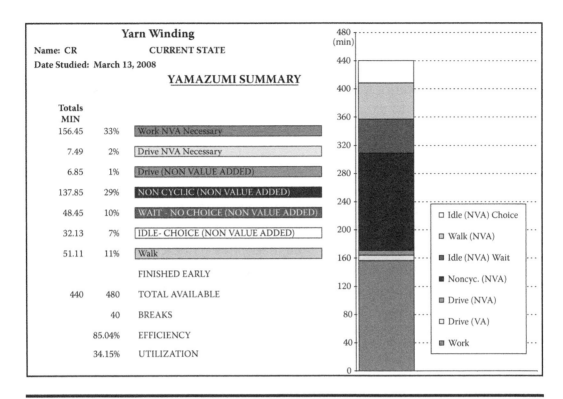

Figure 5.1 Yarn winding individual yamazumi (see the color figure on the CD that accompanies the book).

session of data gathering before continuous improvements are made and sometimes are even higher than we see here.

At this point, for this particular example, it would be a good idea to look at how noncyclic work was defined on the noncyclic Work Summary document (refer to Chapter 4, Figure 4.3). It was defined as Alarm response and Quality checks. Now would be an opportunity to research how many alarm responses, what category of alarm responses, and how much of that 29% of time was dedicated to alarm response. It may help us see if we have a repeatable machine problem. The same should be said of the Quality checks. This may help us identify if we have reoccurring quality problems and help us understand if we have a robust manufacturing process or not. And, look at the Wait No Choice category; 10% of the entire shift was dedicated to Wait No choice. What were the causes? Similarly, 11% of the shift time was spent on Walking. Why was the operator walking? These would all be clues on how to improve this process. You may need to dig deep to thoroughly investigate the root causes of these percentages or they may have been readily observed by the team while collecting data. In almost all cases, the operator can show why these road blocks exist; this is why it is imperative to have them on the team for the entire process review.

We should not overlook the Efficiency and Utilization categories either. I think it is important to examine the definition of Efficiency and Utilization at this particular point.

Efficiency

Efficiency is producing the desired result with a minimum of effort, expense, or waste. For our purpose of NVA, we will use the following formula to calculate our efficiency. All category time, except Idle Choice and Leave Early, divided by the net time available will be our efficiency. This is because these two categories are at the discretion of the operator, not the process. Although several of the other categories are NVA, this is usually inherent to the process, not the operator.

Utilization

Utilization is defined as to make use of. This shows us how well we make use of the time available by adding together the total time of NVA Necessary categories and dividing that number by the net available time.

So let's review our Efficiency and Utilization for this example (Figure 5.1). Our Efficiency is calculated to be 85%. Based on our definition of efficiency, this means that during 85% of the workday there is an activity occurring in this process. However, if we look at the Utilization, it is only 34%. For this particular process, we are only making use of the workday 34% of time in the NVA Necessary categories. This means that we have an opportunity for a 51% improvement.

In Figure 5.2, you can see that the NVA categories of noncyclic and Idle Choice clearly show opportunities for improvements. As I mentioned earlier, the power of the visual graph is clear. In this example, 126.1 minutes or 29% of this operator's available shift time is spent in these two categories. We should rely on the team members to share their observations of what was going on in this category. If we look back at the noncyclic Work Summary sheet (Figure 3.4), noncyclic was defined as building boxes to hold parts as an order is gathered, changing batteries in the scanner or truck itself. These actions encompassed 19% or 82.5 minutes of the entire shift. Knowing this would get me excited about an opportunity for improvement. The category

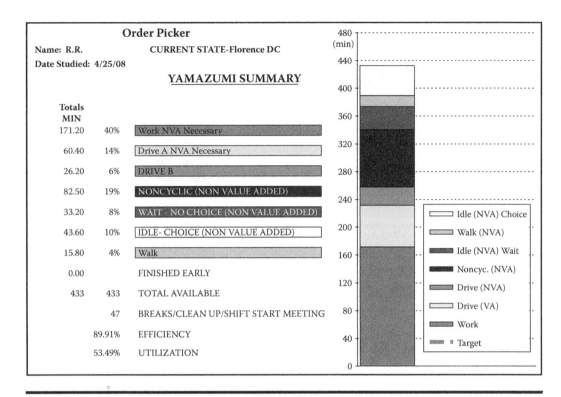

Figure 5.2 Order picker individual yamazumi (see the color figure on the CD that accompanies the book).

of Idle Choice took up 10% of the shift. What were the causes? This may be a clue to a condition of work imbalance. It should be very apparent that this process is broken and needs to be fixed. Our priorities for action planning have already been established and takes out any guess work as to what category to first start reducing/eliminating waste.

In this example, our Efficiency is calculated at almost 90% and our Utilization is calculated at 53%. Again, we have a large opportunity to eliminate waste and focus more time of this process into the areas of NVA Necessary work.

In Figure 5.3, there are categories of NVA Work, noncyclic activities, Idle Choice, and Walk where improvements can be made. Let's review what was defined as NVA work from our noncyclical Work Summary sheet (refer to Figure 4.5). It was defined as "replacing missing operators and running machines." In this case, the role and responsibilities of a team leader should not be running machines. There is an obvious problem that is requiring the team leader to spend so much time in this category and it appears to be that it lies in missing operators.

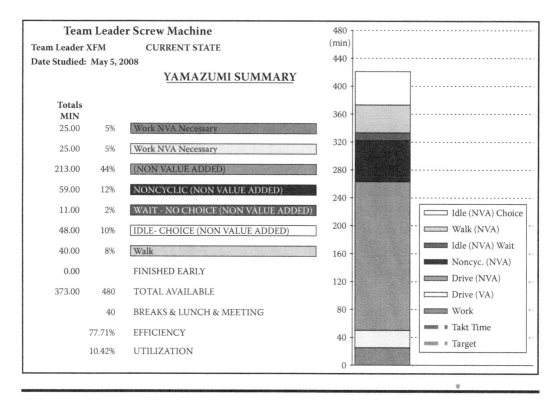

Figure 5.3   Team leader screw machine individual yamazumi (see the color figure on the CD that accompanies the book).

As we review our Efficiency and Utilization numbers, it shows that 77% of the workday is spent in activity in this process. But, with a Utilization number of only 10%, it is clear that the role of the team leader is not being used as intended. The major roadblock for the team leader is that of missing operators. This is either due to a lack of operators or a major attendance problem. Both of these issues are out of the team leader's control. It should be clear as well with this information that the duties of team leader are suffering drastically, which will cause many other problems to other processes because the team leader is running machines instead of attending to his/her main responsibilities. Again, this a problem with the process, not the person.

In Figure 5.4, the yamazumi shows that 58% of the workday is being spent on the main NVA Necessary category. On the noncyclical Work Summary (refer to Figure 4.6), this was defined as planned maintenance activity, job prep, documentation hoist inspections, etc. This is very encouraging. However, we see two opportunities of Idle Choice and Walking. Just these two categories are consuming 125.2 minutes or 26% of the entire

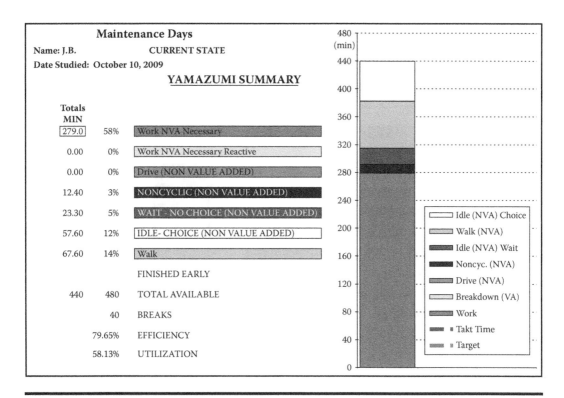

Figure 5.4 Maintenance individual yamazumi (see the color figure on the CD that accompanies the book).

available shift time. Again, this is a big opportunity to fix this problem. The individual yamazumi is helping to show us where the priority of our waste reducing/eliminating and process improvement needs to start.

As we review our Efficiency and Utilization numbers of 80% and 58%, respectively, the data are telling us that we are still underutilizing this position due to the Walking and Idle Choice categories. If we investigate the causes of the Walking (perhaps it was going to get spare parts, for instance), there are solutions available to reduce this dilemma. If we investigate the causes of Idle Choice, we need to find out what in the process is allowing this to occur. If it is a personal choice on the part of the operator, then this becomes a management issue that needs to be addressed. This can be accomplished with an explanation of the expectations of the position and the education as to the impact on the process this personal choice is causing. We should ask the team members who collected the data to share their observations of what were the causes of the problems in these two categories.

In Figure 5.5, the good news is that 49% of the time is being spent in the NVA Necessary categories, the bad news is that 34% of the time is

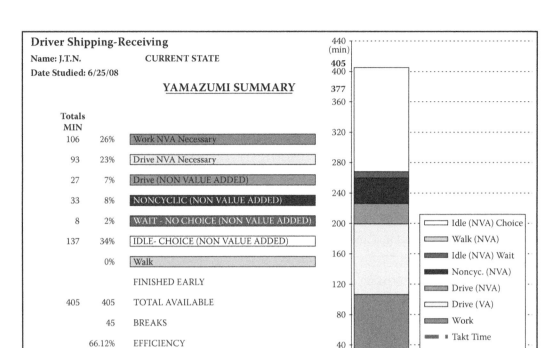

Figure 5.5 Fork truck driver individual yamazumi (see the color figure on the CD that accompanies the book).

being spent in the Idle Choice category. This, of course, will have a major impact on our Efficiency and Utilization numbers. Also, we need to be cautious and not make a judgment until we find the cause or causes. At first glance, it may appear that this block of time is totally due to operator discretion. Fork truck drivers' process is generally reviewed as part of a group. If this were true in this case, an additional piece of information that could prove helpful would be to look at the group yamazumi for this particular process. Don't forget that we should also ask the team members, who collected the data and observed the process first hand, what they saw. It should be clear, though, that this process has ample opportunity to eliminate waste.

So, let's review our Efficiency and Utilization numbers.. Our Efficiency is calculated to be 66%. This has been negatively impacted due to the large amount of time in the Idle Choice category. Our Utilization is calculated as 49% again due to the large block of time in the Idle Choice category. As a reminder, we need to get to the root cause of why so much time was Idle Choice and then fix the process.

The Group Graph

Once the individual information has been populated, either by manual or scanned methods, the Group graph will automatically be created. This is a fantastic visual display of all your individual information pulled together on one page. You will undoubtedly see variation between people doing the same process. Variation is the root of many problems in any process, whether it's productivity or quality dimensions. With the group yamazumi, you really get a clear picture of how the entire process is functioning.

Completed Examples

Let's look again at some examples, and I think you will find them very enlightening.

You can see quite a bit of variation between the operators in Figure 5.6. Operators 1 and 2 are similar, but there is a large gap between these positions and the others. There is plenty of opportunity displayed here in at least four areas. These four areas are Work Balancing, Walking, Waiting, and Idle Choice As stated earlier, with the group yamazumi, you really get a clear picture of how the entire process is functioning.

In Figure 5.7, we again see variation between the pickers within the same activity. The categories that would have a priority to focus on would be noncyclic activities, Drive B, and the Wait No Choice. Based on some past experience, I would be pretty sure that this location is probably behind on customer orders, working some overtime, and most likely experiencing premium shipments due to these three categories. The variation needs to be eliminated and the work balanced. Once this is done, it will be easier to make and sustain changes.

There is definitely an issue to be addressed in Figure 5.8 that is preventing all the team leaders from performing the desired functions of their jobs. Look at the disproportionate time of the NVA category compared to all the others. During the job category matrix review, it is clear that replacing missing operators was a scope of activity we identified as total NVA to this position, but that is exactly where the team leaders are spending the majority of their time. This issue isn't about work balance, it is about what actions they are performing. The cross-functional team in this situation has clearly a need to perform a root cause analysis of why this is occurring and put a process in place to stop it, if they ever want to get the value out of the team leader position.

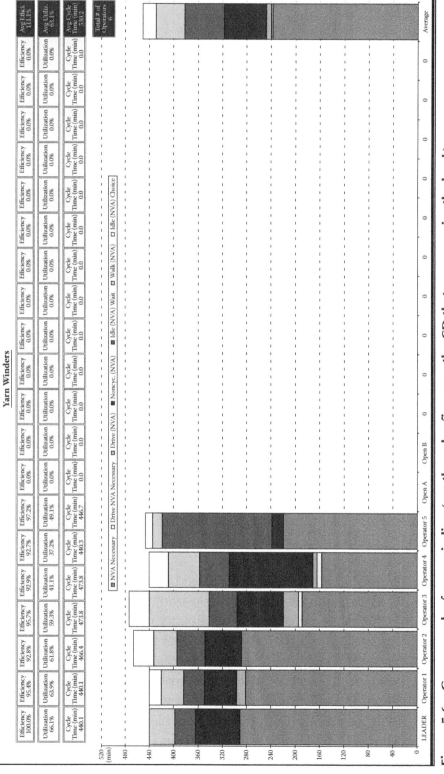

Figure 5.6 Group graph of yarn winding (see the color figure on the CD that accompanies the book).

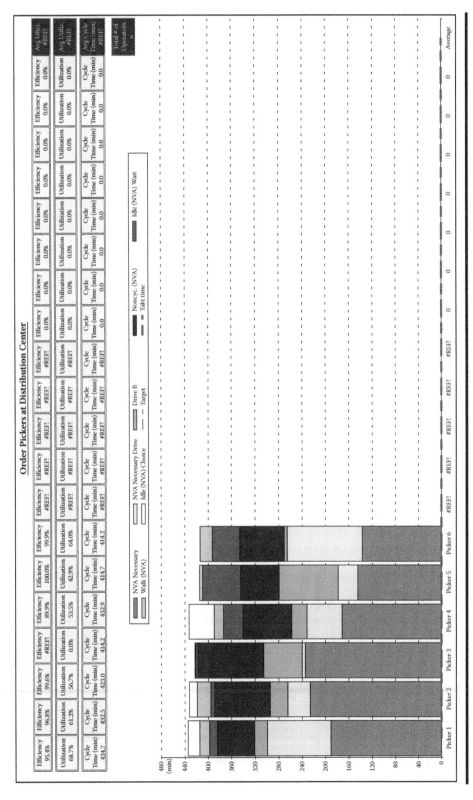

Figure 5.7   Group graph of order picker (see the color figure on the CD that accompanies the book).

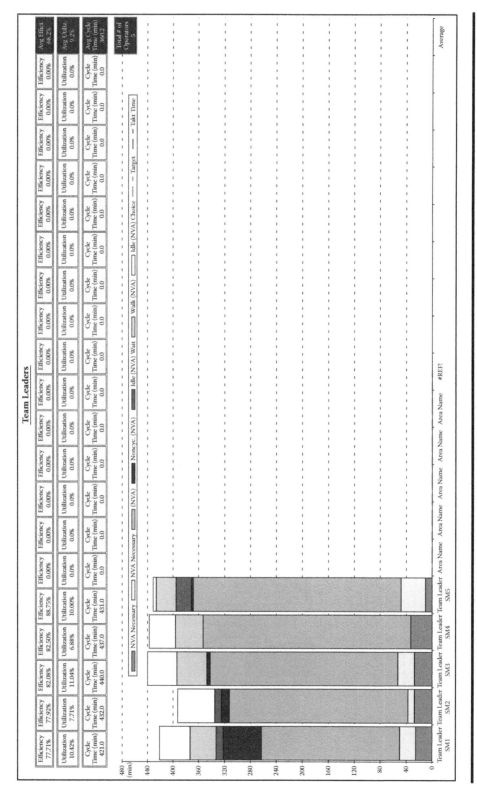

Figure 5.8   Group graph of team leader screw machines (see the color figure on the CD that accompanies the book).

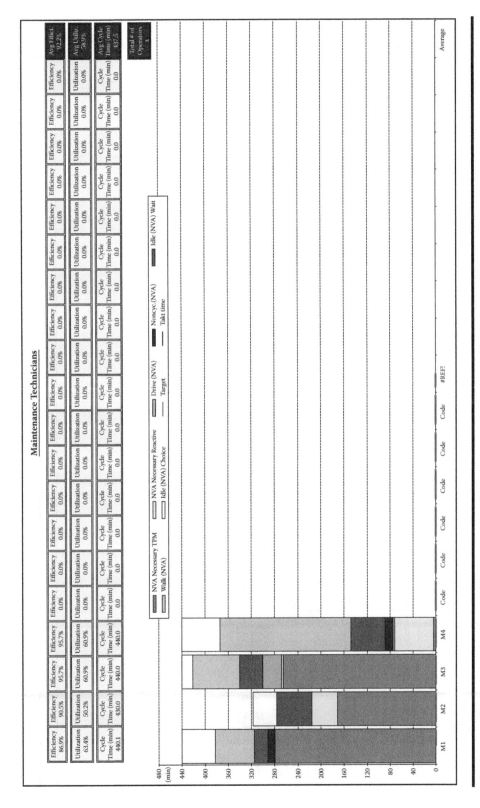

Figure 5.9 Group graph of maintenance (see the color figure on the CD that accompanies the book).

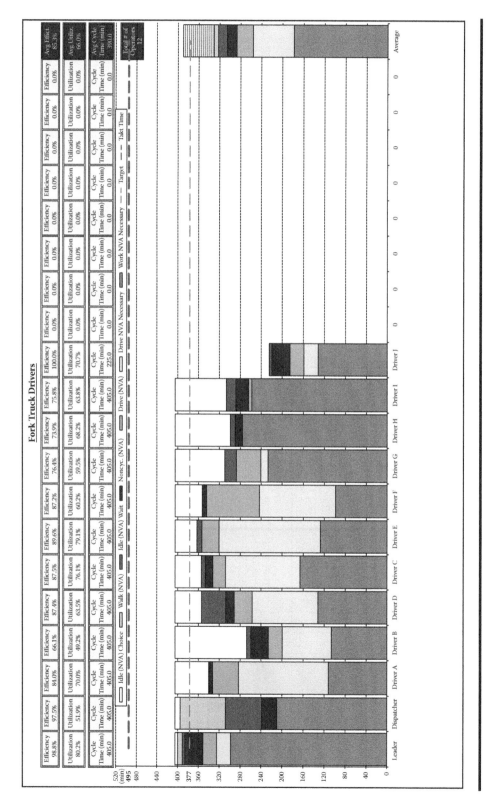

Figure 5.10 Group graph of fork truck driver (see the color figure on the CD that accompanies the book).

In Figure 5.9, we see a perfect example of why we use all the tools in developing standardized work for noncyclical processes. If we had simply looked at the individual yamazumis, I do not think we would have seen the impact Walking has on the entire utilization of the maintenance technicians. You can see one of the reasons why the color coding for the various NVA Necessary and NVA titles is helpful. This is a great graphical representation of what the current state of this maintenance group is. What two areas would you choose as a priority on which to focus improvements? I hope you selected Walking and Wait No Choice. Based on the areas of opportunity displayed and once we know what is causing these conditions, there are further Lean tools that can be instituted to help this group perform better; not harder, but smarter.

The graph in Figure 5.10 shows exactly why fork truck drivers' jobs are generally an excellent opportunity for improvement. Look at the variation from driver to driver. Normally, even the people performing the jobs are not aware of the disparities displayed, such as this one. Not only is there an opportunity to rebalance the work, but there also may be an opportunity to redeploy one driver to another area of need, based on the Idle Choice time that is displayed in the Group graph.

Chapter 6

Sustaining Improvements and Tools for Making Improvements

Because decisions or changes should never be made without data, you are now in an excellent position to do both. You have individual and group yamazumi charts that clearly show you the gaps to 85% utilization and you have the noncyclic Work Summary sheets that break down the data into finer details. It is now time for the cross-functional team to start brainstorming ideas that will achieve that goal. This is why it is very important for the team leader of this exercise to be a strong facilitator and also have a strong knowledge in the various tools of Lean production.

You must stabilize, standardize, and eliminate the variation to make continuous improvements. So, first concentrate on balancing the work between the group members. Think of the big picture with the information you have. You will need to create and/or modify work instructions and Standard Work charts to help in this matter.

Help the cross-functional team members think of the big picture with the information you have. For example, if your individual yamazumi shows a person walking 67 minutes, you can calculate what impact that has over the long view. For example purposes, let's say there are 255 work days a year: 255 work days × 67 minutes/day = 17, 085 minutes a year of walking divided by 60 minutes = 284.75 hours/year divided by 40 hours = 7.12 weeks a year of just walking. This is only one example, but this is a view that has to be considered. Again, most people don't realize the impact on the big picture, so I think it is worth the time to share and show everyone involved. In this chapter, in addition to speaking to sustainment, I would also like to share with you some of the tools that were used to make continuous improvements for some of the conditions cited in the other chapters.

Sustaining

Layered Audits: Sustaining

Once you make improvements, it is important that you are able to sustain these improvements. To sustain your improvements, you must put auditing and countermeasures in place (Figure 6.1). If you do not, your changes are sure to fail.

There are four levels of auditing that must take place:

1. Each shift: This is usually done by the team or group leader of the shift.
2. Daily: This is usually done by the supervisor choosing a position randomly once a day.
3. Weekly: This is usually done by the general supervisor or to whom the supervisor reports.
4. Monthly: This is usually done by the plant manager or a department head.

As you can see, this requires a commitment and understanding of what should be occurring on the *Gemba* (the actual place) by all levels. The countermeasures must be such that if any deviation or abnormality is observed during the audit, the process must be put back into a normal or standard condition immediately. The bottom line to sustainment as a Lean leader is a "zero tolerance" for out-of-standard conditions whether observed during an audit or any time it is observed.

Tools for Creating Continuous Improvements

My Toyota sensei taught me that generally you need to make continuous improvements to a process about nine times before it will cost you more money than you will gain from an improvement exercise; sometimes it could be more depending on how much and how fast you can implement and, more importantly, sustain improvement. One important point to remember here is that you must stabilize your group of improvements for a proven

DAILY STANDARDIZED WORK AUDIT

Dept-Area Audited

Month of Audit

Auditor

Date of Audit

1	Are the standard work sheets visible?
2	Are the walk patterns being followed?
3	Is work being performed to the target cycle time?
4	Is the constraint being run properly?
5	Are the work instructions being followed?
6	Are parts presentation devices in their proper location?
7	Are the metric and/or targets updated to the posted plan?
8	Is the area reporting board being updated?
9	Do the downtime reasons explain total gap (lost pcs)?
10	Is Mgt. auditing the production reporting boards?

Comments

Audit Procedure

1 Check each cell, once per day
2 Observe straight cycles
3 If an item is in conformance,
   place a "O" in the block
4 If an item is not in conformance,
   place a "X" in the block & list the root cause &
   countermeasure in the comments section.

Figure 6.1   Sustaining audit.

period of time before you move to *kaizen* the process again. (Kaizen is Japanese for "continuous improvement.")

Point of Use Cribs

Once we had collected the data on our noncyclic Work Summary sheets (refer to Figure 4.6) for the maintenance technicians group, it was clear that nonvalue add (NVA) Walking was a top Pareto (a visual tool used to identify the causes of problems that occur most frequently) cause having a negative impact on nonvalue add but necessary (NVA Necessary); this was also verified by the Group yamazumi (refer to Figure 5.9). Upon further investigation, it was discovered that the bulk of the NVA Walking was to the MRO (maintenance repair and operations, or spare parts) crib located at the opposite end of the plant. To help reduce this situation, the concept of the Point of Use cribs were implemented and installed throughout the facility.

The key features of a Point of Use (POU) crib is to have high-usage items readily accessible (this is based on data from your Computerized Maintenance Management System [CMMS]), self-accessed by the associates qualified to do the work, accessed without paperwork, and kept properly stocked using using a pull system with kanban cards.

There are eight steps to the POU process:

1. Determine the number and location(s).
   - Review the plant address system.
   - Review existing POU area(s) (general stores, crib, satellite cribs).
   - Determine where new POU area(s) should be set up; coordinate with manufacturing, maintenance, and purchasing.
   - Interview potential customers (maintenance technicians) for possible locations, suggestions of parts that may be part of the "tribal knowledge" of the area.

2. Establish the spare parts and quantities for each POU.
   - Analyze the parts and quantities for each POU using historical repair and breakdown data.
   - Determine a method to add or delete parts in the POU area.

3. Determine the resource(s) and designate pull method for parts replenishment.
   - Select a pull method(s) to use within your POU area.

4. Develop visual controls.
   - Develop visual controls that you will use within your POU area.

5. Identify and install equipment.
   - Select types and sizes of bins, racks, cabinets, and storyboards.
   - Install equipment.

6. Label bins and stock the POU.
   - Create the label that will be used in the POU areas.
   - Label and stock the bins, cabinets, and racks.

7. Determine internal delivery requirements.
   - Review existing delivery routes.
   - Establish delivery addresses.
   - Develop new scheduled delivery routes.
   - Develop procedures or job instructions for delivery drivers.
   - Communicate delivery procedures to supplier.
   - Develop standards for delivery.

8. Create a plan to train the end users.
   - Create a plan to train everyone attached to the process.
   - Understand that the maintenance technician is the customer of this process.

Some of the expected benefits to implementing and sustaining POU cribs are:

■ Reduced walking time for maintenance and technicians
■ Will help clean and organize areas
■ Reduced downtime through improved part availability
■ Improved spare parts inventory status by reducing/eliminating spare parts "stashes"
■ Involve maintenance and technicians in the spare parts process

Implementation success factors include:

■ POU parameters: "Lead Time versus Need Time"
■ POU location(s)
■ Inventory item sheets
■ Resource allocation: Replenishment and inventory scrubbing

- Identify routing
- Identify type of POU equipment: Containers, tables, vidmars, etc.
- Establish preliminary timeline
- Inventory level: Tracking POU parts location and usage
- Document POU process: "Before and after" pictures
- Secure top leadership commitment and secure active involvement of maintenance and technicians
- Document and broadcast successes

If you determine that using PULL cards is the best method, Figure 6.2 and Figure 6.3 will be of some help in developing your PULL system. Figure 6.2 is an example of a PULL card that should be kept with the part. Figure 6.3 is an example of a PULL card permanently attached to the cabinet at the point where the part is stored.

There are many types of POU cribs to implement. The important item to remember is to use the simple methods first. They require less attention and cost the least. I have attached some additional examples that you may find helpful (Figure 6.4 to Figure 6.9).

Point of Use PULL Card for a Part

The Pull Card will be kept with the part in the cabinet.

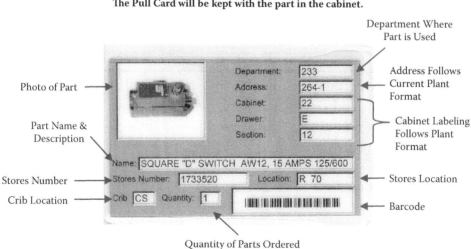

Card is sized to fit a standard drivers license size laminating pouch.

Figure 6.2   An example of a PULL card that should be kept with the part.

Point of Use PULL Card for Part Location

A card will be permanently attached to the cabinet at the point where the part is stored.

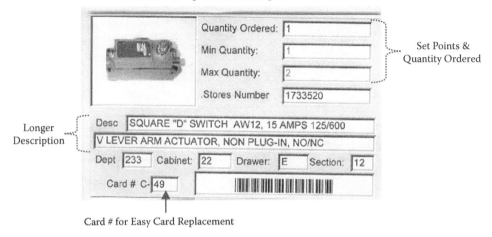

Set Points & Quantity Ordered

Longer Description

Card # for Easy Card Replacement

Figure 6.3   An example of a PULL card permanently attached to the cabinet at the point where the part is stored.

Bins on a Shelf

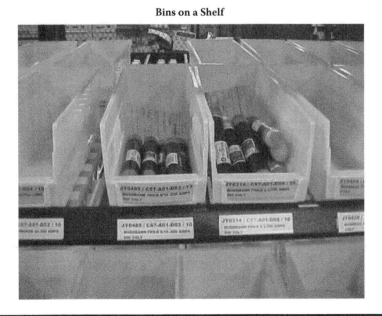

Figure 6.4   PULL card for part

Divided Shelves

Figure 6.5   PULL card for part location

Lazy Susan

Figure 6.6   Bins on a shelf

Bins Attached to a Rack

Figure 6.7 Divided shelves

Vending Machines

Figure 6.8 Lazy susan

Cabinets

Figure 6.9   Bins attached to a rack

Shipping and Receiving Windows

As we saw in the fork truck drivers Group yamazumi (refer to Figure 5.10), we saw that there were several categories that were having a negative impact on their daily activities. These specifically were noncyclic activities, Idle, and Wait. On further examination, in the instant case, what was found was that the activities most desired by this group, shipping and receiving, were not structured. One tool that was implemented, Shipping and Receiving Windows, provided a more stabilized workday for the entire group, therefore, providing an environment for continuous improvement. By using very specific shipping windows, you can manage and predict time to allocate for this activity. Figure 6.10 is an example of Shipping Window Boards.

The same is true for the Receiving process. By using a planned process and timing for receiving supplier goods, you will be able to manage the associates time toward more NVA Necessary work, thus, getting a bigger bang for your buck, so to speak (Figure 6.11).

Route Maps, Pick Up, Delivery Instructions, and Visual Aids

Another group of tools that can be utilized to eliminate waste for a fork truck driver or material handler is Route Maps that include location,

Ship To Schedule

Scheduled Window Shipping

☞ Scheduled Window Shipping is a specific dock schedule allocating days and times of day for delivery of empty finished goods containers and pick-up of finished goods (Shipping to Schedule).

Figure 6.10   Example of shipping window boards.

Supplier Delivery

Scheduled Windows Receiving

☞ Scheduled Window Receiving is a specific dock schedule allocating days and times of day for delivery of purchased material and pick-up of empty/returnable containers.

Figure 6.11   Example of receiving window boards.

delivery, and pick up information as well as the timing of the route. This is important as there may be several routes that a fork truck driver or material handler cover in a shift. When implementing designated routes, don't forget to incorporate visual aids in the route (Figure 6.12 to Figure 6.15).

Instruction #:

Instruction Title: STOCK ROUTE - Gilmag

FS-143

Ind. Engineer	Supervisor	Prepared by:		Date (Original):	8/1/2008
J.M	J.N	J. Moon / J. Noon		Date (Revision):	12/15/2009
12/15/09	12/15/09	Mail Stop: AB - 12		Line Rate (pcs/hr):	600
		Phone # : BR - 549		Route Cycle Time (sec):	360

WORKPLACE LAYOUT

# of Tote Lifts >10 lbs.	Weight of One Full Tote (lbs)
20	negligible
4	30
4	12
4	14
8	24
5	17
4	27
1	9
8	35
8	35
66	

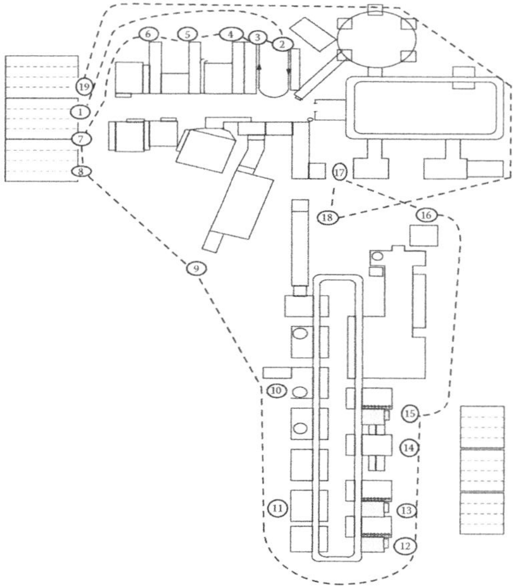

Figure 6.12 Stock route example.

Flex Auto #4 (FA-4) Materials				
Stop Number	Part Description	P/U Point	Part Number(s)	Method
1A	Piston Assy.	G-45	04992422	Place buckhorn onto creform rack,
				pick up empty(s) and pull cards.
1B	Piston Plate Assy.	P28	22172498	Place buckhorn onto creform rack,
			22140825	pick up empty(s) and pull cards.
			22140788	" " " "
			22140804	" " " "
			22187362	" " " "
			22187356	" " " "
			22187358	" " " "
			22187360	" " " "
			22187763	" " " "
2	Rod Guide	ABB51	22172144	Place cardborad box onto creform rack,
				pick up pull cards & dispose of empty box(es)
3	Top Mounting Stud	K77	22177751	Place tote and/or cardboard box onto rack,
	Seal Cover	K77	22050280	pick up empty(s), pick up pull cards,
	Seal	ABB56	22122908	dispose of empty cardboard box.
	Seal Spring	AW71	22182731	" " " "
	Top Mounting Ring	K77	22148845	" " " "
4	Cover Plate	K77	22120657	Place tote onto creform rack,
				pick up empty(s) and pull cards.
	Piston Rods	K49	22125009	Tug cart to cell, place in designated area,
			22125027	tug away empty cart.
5	Reservoir Tubes	D56	04992338	Place tote onto creform rack,
			04992355	pick up empty(s) and pull cards.
6	Base Cup	K77	22104391	Place tote onto creform rack,
				pick up empty(s) and pull cards.
7	Cylinder Tube	E56	22152298	Place tote onto creform rack,
			22152197	pick up empty(s) and pull cards.
8	Compression Valves	AX34	22140774	Place tote onto creform rack,
			22140946	pick up empty(s) and pull cards.
			22140780	" " " "
			22140913	" " " "
			22187371	" " " "
			22187374	" " " "
			22187760	" " " "
8B	Bottom Ring	K77	22148851	" " " "

Figure 6.13   Delivery instructions example.

Tugger Route Flex Cell #4 Materials				
Stop Number	Part Description	P/U Point	Part Number(s)	Method
1	Flex #4 Cell	Cell	As needed	Using cell delivery and pull card instructions, service the cell area.
2	Compression Valves	AX34	22140774	Follow instructions on back of pull card.
			22140946	
			22140780	
			22140913	
			22187371	
			22187374	
			22187760	
3	Piston Plate Assy.	P28	22172498	Follow instructions on back of pull card.
			22140825	
			22140788	
			22140804	
			22187362	
			22187356	
			22187358	
			22187360	
			22187763	
4	Piston Paks	G45	04992422	Follow instructions on back of pull card.
5	Piston Rods	K49	22125009	Follow instructions on back of pull card.
			22125209	
6	Top Mounting Stud	K77	22177751	Follow instructions on back of pull card.
	Seal Cover	K77	22050280	
	Top Mounting Ring	K77	22148845	
	Cover Plate	K77	22120657	
	Base Cup	K77	22104391	
	Bottom Ring	K77	22148851	
7	Seal Spring	AW71	22182731	Follow instructions on back of pull card.
8	Rod Guide	ABB51	22172144	Follow instructions on back of pull card.
9	Cylinder Tube	D56	22152298	Follow instructions on back of pull card.
			22152197	
	Reservoir Tube	E56	04992338	Follow instructions on back of pull card.
			04992355	
10	Seal	ABB56	22122908	Follow instructions on back of pull card.

Figure 6.14   Pick up instructions example.

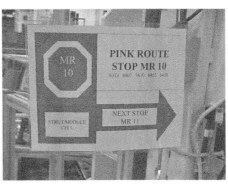

Figure 6.15　Stop sign examples.

These are all methods to eliminate waste to this noncyclical process that have been proven to work.

I hope you have enjoyed this material and, even more, are applying it to the noncyclical process in your facility. If you follow this process, you will be successful, improve the productivity, and increase the profits of your organization.

Index

About the Author

Joseph Niederstadt was born in Saginaw, Michigan, in the early 1950s. This was a town that depended heavily on the automotive industry, as did many midwest towns during this time period. Industry was booming and mass production was well on the way. The automotive industry employed tens of thousands in the area, providing a standard of living for those working in the factories that few had ever seen or probably will ever see again.

Like many others in the area, Niederstadt started work in a General Motors factory with the goal of making money and providing benefits for a family that was yet to come. He never realized that it was the beginning of a career in manufacturing that would span beyond 30 years. His first experience was working in one of GM's foundries located in the area. His many jobs included:

- Trim deck operator (loading the furnace charging buckets with 7.5 tons of mixed parts to be melted).
- Overhead crane operator (consisted of driving a crane with a bucket of several tons of molten iron) on an overhead rail and delivering small amounts of molten iron to other locations.
- Furnace tear out crew, where the job was to rebuild the lining of an electric induction furnace 20 feet from the surface using a 90-pound jackhammer in 100+ degree heat (many times the wood forms he was standing on would start on fire due to the heat).

Later at the Chevrolet Motor Division, Niederstadt was an assembly line operator, a material handler for the assembly line, repair man for assemblies, and a part-time fill-in for the assembly job setter. After several years in these positions, he applied for and was accepted for a supervisor's role. Whether it was at GM or Delphi Automotive Systems, his supervisory experiences included broaching, machining, assembly, after-market operations, quality control, maintenance, receiving, shipping, production control and logistics (including coordination of the entire facilities annual inventory activities), stampings, blanking, rubber extrusion, and plastic injection molding. Later, as advancements came his way, he also gained experience in labor relations, implementing, coaching and mentoring others through the Divisional Lean Core Team, supplier development, and international assignments. He also won one of the very first mentoring awards given by the Delphi Corporation (Troy, Michigan). During this progression, Niederstadt has never forgotten his roots as an operator and has always strived to make the work environment better for the operator.

He has lived through the transition from mass to lean production, from the "do as I tell you" mentality to a team-managed work system, from massive inventories to "Just In Time" philosophies, from "run all you can every time you can" thinking to PULL systems, from dedicated equipment to flexible cells, from changeovers that took days to ones that now take only minutes … the list goes on and on. He has been taught by several senseis from Toyota as well as some of the best Lean people at GM and Delphi.

Niederstadt has worked in the United States, Canada, Mexico, Brazil, Taiwan, India, Korea, China, and Thailand, applying the system approach that is included in this book. Most recently, he was the director of Lean Applications Asia–Pacific Region for a global corporation and responsible for leading the implementation of Lean in 26 facilities in 11 different countries. He is currently an independent consultant applying Lean tools to manufacturing and SGA (salary general and administrative) operations.

Printed and bound by CPI Group (UK) Ltd, Croydon, CR0 4YY

23/10/2024

01777685-0011